2005 EDITION

ASD/LRFD

MANUAL

FOR ENGINEERED
WOOD CONSTRUCTION

ASD/LRFD Manual for Engineered Wood Construction 2005 Edition

First Printing: August 2006
Second Printing: February 2007
Third Printing: February 2008

ISBN 0-9625985-7-7 (Volume 3)
ISBN 0-9625985-8-5 (4 Volume Set)

Copyright Permission
AF&PA American Wood Council
1111 Nineteenth St., NW, Suite 800
Washington, DC 20036
email: awcinfo@afandpa.org

Printed in the United States of America

OREWORD

This *Allowable Stress Design/Load and Resistance actor Design Manual for Engineered Wood Construction SD/LRFD Manual)* provides guidance for design of most ood-based structural products used in the construction wood buildings. The complete *Wood Design Package* cludes this *ASD/LRFD Manual* and the following:

- *ANSI/AF&PA NDS-2005 National Design Specification® (NDS®) for Wood Construction* – with Commentary; and, *NDS Supplement – Design Values for Wood Construction, 2005 Edition,*
- *ANSI/AF&PA SDPWS-05 – Special Design Provisions for Wind and Seismic (SDPWS)* – with Commentary,
- *ASD/LRFD Structural Wood Design Solved Example Problems, 2005 Edition.*

The American Forest & Paper Association (AF&PA) s developed this manual for design professionals. F&PA and its predecessor organizations have provided gineering design information to users of structural wood products for over 70 years, first in the form of the *Wood Structural Design Data* series and then in the *National Design Specification (NDS) for Wood Construction.*

It is intended that this document be used in conjunction with competent engineering design, accurate fabrication, and adequate supervision of construction. AF&PA does not assume any responsibility for errors or omissions in the document, nor for engineering designs, plans, or construction prepared from it.

Those using this standard assume all liability arising from its use. The design of engineered structures is within the scope of expertise of licensed engineers, architects, or other licensed professionals for applications to a particular structure.

American Forest & Paper Association

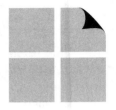

TABLE OF CONTENTS

Part/Title   Page

M1 GENERAL REQUIREMENTS FOR STRUCTURAL DESIGN 1
M1.1 Products Covered in This Manual
M1.2 General Requirements
M1.3 Design Procedures

M2 DESIGN VALUES FOR STRUCTURAL MEMBERS 3
M2.1 General Information
M2.2 Reference Design Values
M2.3 Adjustment of Design Values

M3 DESIGN PROVISIONS AND EQUATIONS 5
M3.1 General
M3.2 Bending Members - General
M3.3 Bending Members - Flexure
M3.4 Bending Members - Shear
M3.5 Bending Members - Deflection
M3.6 Compression Members
M3.7 Solid Columns
M3.8 Tension Members
M3.9 Combined Bending and Axial Loading
M3.10 Design for Bearing

M4 SAWN LUMBER 11
M4.1 General
M4.2 Reference Design Values
M4.3 Adjustment of Reference Design Values
M4.4 Special Design Considerations
M4.5 Member Selection Tables
M4.6 Examples of Capacity Table Development

M5 STRUCTURAL GLUED LAMINATED TIMBER 27
M5.1 General
M5.2 Reference Design Values
M5.3 Adjustment of Reference Design Values
M5.4 Special Design Considerations

M6 ROUND TIMBER POLES AND PILES 33
M6.1 General
M6.2 Reference Design Values
M6.3 Adjustment of Reference Desi[gn] Values
M6.4 Special Design Consideration

M7 PREFABRICATED WOOD I-JOISTS 37
M7.1 General
M7.2 Reference Design Values
M7.3 Adjustment of Reference Design Values
M7.4 Special Design Considerations

M8 STRUCTURAL COMPOSITE LUMBER 53
M8.1 General
M8.2 Reference Design Values
M8.3 Adjustment of Reference Design Values
M8.4 Special Design Considerations

M9 WOOD STRUCTURAL PANELS 59
M9.1 General
M9.2 Reference Design Values
M9.3 Adjustment of Reference Design Values
M9.4 Special Design Considerations

M10 MECHANICAL CONNECTIONS 69
M10.1 General
M10.2 Reference Design Values
M10.3 Design Adjustment Factors
M10.4 Typical Connection Details
M10.5 Pre-Engineered Metal Connectors

M11 DOWEL-TYPE FASTENERS 85
M11.1 General
M11.2 Reference Withdrawal Design Values
M11.3 Reference Lateral Design Values
M11.4 Combined Lateral and Withdrawal Loads
M11.5 Adjustment of Reference Design Values
M11.6 Multiple Fasteners

M12 SPLIT RING AND SHEAR PLATE CONNECTORS 89
General
Reference Design Values
Placement of Split Ring and Shear [Pla]te Connectors

Part/Title   Page

M13 TIMBER RIVETS.............................91
M13.1 General
M13.2 Reference Design Values
M13.3 Placement of Timber Rivets

M14 SHEAR WALLS AND DIAPHRAGMS.............................93
M14.1 General
M14.2 Design Principles
M14.3 Shear Walls
M14.4 Diaphragms

M15 SPECIAL LOADING CONDITIONS.............................99
M15.1 Lateral Distribution of Concentrated Loads
M15.2 Spaced Columns
M15.3 Built-Up Columns
M15.4 Wood Columns with Side Loads and Eccentricity

Part/Title   Pag

M16 FIRE DESIGN.............................101
M16.1 General
M16.2 Design Procedures for Exposed Wood Members
M16.3 Wood Connections

LIST OF TABLES

M4.3-1   Applicability of Adjustment Factors for Sawn Lumber 13

M4.4-1   Approximate Moisture and Thermal Dimensional Changes............................... 14

M4.4-2   Coefficient of Moisture Expansion, e_{ME}, and Fiber Saturation Point, FSP, for Solid Woods.. 15

M4.4-3   Coefficient of Thermal Expansion, e_{TE}, for Solid Woods 16

M4.5-1a   ASD Tension Member Capacity (T'), Structural Lumber (2-inch nominal thickness Visually Graded Lumber (1.5 inch dry dressed size), $C_D = 1.0.4$-inch nominal thickness Visually Graded Lumber (3.5 inch dry dressed size), $C_D = 1.0$) 18

M4.5-1b   ASD Tension Member Capacity (T'), Structural Lumber (2-inch nominal thickness MSR Lumber (1.5 inch dry dressed size), $C_D = 1.0$) 18

M4.5-2a   ASD Column Capacity (P', P'_x, P'_y), Timbers (6-inch nominal thickness (5.5 inch dry dressed size), $C_D = 1.0$) 19

M4.5-2b   ASD Column Capacity (P', P'_x, P'_y), Timbers (8-inch nominal thickness (7.5 inch dry dressed size), $C_D = 1.0$) 20

M4.5-2c   ASD Column Capacity (P', P'_x, P'_y), Timbers (10-inch nominal thickness (9.5 inch dry dressed size), $C_D = 1.0$) 21

M4.5-3a   ASD Bending Member Capacity (M', C_rM', V', and EI), Structural Lumber (2-inch nominal thickness (1.5 inch dry dressed size), $C_D = 1.0$, $C_L = 1.0$) 22

M4.5-3b   ASD Bending Member Capacity (M', C_rM', V', and EI), Structural Lumber (4-inch nominal thickness (3.5 inch dry dressed size), $C_D = 1.0$, $C_L = 1.0$. 22

M4.5-4a   ASD Bending Member Capacity (M', V', and EI), Timbers (6-inch nominal thickness (5.5 inch dry dressed size), $C_D = 1.0$, $C_L = 1.0$) 23

M4.5-4b   ASD Bending Member Capacity (M', V', and EI), Timbers (8-inch nominal thickness (7.5 inch dry dressed size), $C_D = 1.0$, $C_L = 1.0$)..................................... 23

M4.5-4c   ASD Bending Member Capacity (M', V', and EI), Timbers (10-inch nominal thickness (9.5 inch dry dressed size), $C_D = 1.0$, $C_L = 1.0$)..................................... 24

M4.5-4d   ASD Bending Member Capacity (M', V', and EI), Timbers (Nominal dimensions > 10 inch (actual = nominal – 1/2 inch), $C_D = 1.0$, $C_L = 1.0$)..................................... 24

M5.1-1   Economical Spans for Structural Glued Laminated Timber Framing Systems 29

M5.3-1   Applicability of Adjustment Factors for Structural Glued Laminated Timber........... 31

M5.4-1   Average Specific Gravity and Weight Factor 32

M6.3-1   Applicability of Adjustment Factors for Round Timber Poles and Piles 35

M7.3-1   Applicability of Adjustment Factors for Prefabricated Wood I-Joists 40

M8.3-1   Applicability of Adjustment Factors for Structural Composite Lumber.................... 56

M9.1-1   Guide to Panel Use................................... 61

M9.2-1   Wood Structural Panel Bending Stiffness and Strength 62

M9.2-2   Wood Structural Panel Axial Stiffness, Tension, and Compression Capacities........ 63

M9.2-3   Wood Structural Panel Planar (Rolling) Shear Capacities 65

M9.2-4   Wood Structural Panel Rigidity and Through-the-Thickness Shear Capacities .. 65

M9.3-1   Applicability of Adjustment Factors for Wood Structural Panels 66

M9.4-1   Panel Edge Support................................... 67

M9.4-2   Minimum Nailing for Wood Structural Panel Applications..................................... 68

M10.3-1   Applicability of Adjustment Factors for Mechanical Connections 71

M11.3-1 Applicability of Adjustment Factors for Dowel-Type Fasteners 87

M12.2-1 Applicability of Adjustment Factors for Split Ring and Shear Plate Connectors 90

M13.2-1 Applicability of Adjustment Factors for Timber Rivets 92

M16.1-1 Minimum Sizes to Qualify as Heavy Timber Construction 102

M16.1-2 One-Hour Fire-Rated Load-Bearing Wood-Frame Wall Assemblies 103

M16.1-3 Two-Hour Fire-Rated Load-Bearing Wood-Frame Wall Assemblies 103

M16.1-4 One-Hour Fire-Rated Wood Floor/Ceiling Assemblies 104

M16.1-5 Two-Hour Fire-Rated Wood Floor/Ceiling Assemblies 104

M16.1-6 Minimum Depths at Which Selected Beam Sizes Can Be Adopted for One-Hour Fire Ratings 119

M16.1-7 Fire-Resistive Wood I-Joist Floor/Ceiling Assemblies 123

M16.1-8 Privacy Afforded According to STC Rating 143

M16.1-9 Contributions of Various Products to STC or IIC Rating 143

M16.1-10 Example Calculation 144

M16.1-11 STC & IIC Ratings for UL L528/L529 144

M16.1-12 STC & IIC Ratings for FC-214 144

M16.2-1 Design Load Ratios for Bending Members Exposed on Three Sides (Structural Calculations at Standard Reference Conditions: $C_D = 1.0$, $C_M = 1.0$, $C_t = 1.0$, $C_i = 1.0$, $C_L = 1.0$) (Protected Surface in Depth Direction) 147

M16.2-2 Design Load Ratios for Bending Members Exposed on Four Sides (Structural Calculations at Standard Reference Conditions: $C_D = 1.0$, $C_M = 1.0$, $C_t = 1.0$, $C_i = 1.0$, $C_L = 1.0$) 148

M16.2-3 Design Load Ratios for Compression Members Exposed on Three Sides (Structural Calculations at Standard Reference Conditions: $C_M = 1.0$, $C_t = 1.0$, $C_i = 1.0$) (Protected Surface in Depth Direction) .... 14(

M16.2-4 Design Load Ratios for Compression Members Exposed on Three Sides (Structural Calculations at Standard Reference Conditions: $C_M = 1.0$, $C_t = 1.0$, $C_i = 1.0$) (Protected Surface in Width Direction) .... 15(

M16.2-5 Design Load Ratios for Compression Members Exposed on Four Sides (Structural Calculations at Standard Reference Conditions: $C_M = 1.0$, $C_t = 1.0$, $C_i = 1.0$) .... 15)

M16.2-6 Design Load Ratios for Tension Members Exposed on Three Sides (Structural Calculations at Standard Reference Conditions: $C_D = 1.0$, $C_M = 1.0$, $C_t = 1.0$, $C_i = 1.0$) (Protected Surface in Depth Direction) .... 15?

M16.2-7 Design Load Ratios for Tension Members Exposed on Three Sides (Structural Calculations at Standard Reference Conditions: $C_D = 1.0$, $C_M = 1.0$, $C_t = 1.0$, $C_i = 1.0$) (Protected Surface in Width Direction) 153

M16.2-8 Design Load Ratios for Tension Members Exposed on Four Sides (Structural Calculations at Standard Reference Conditions: $C_D = 1.0$, $C_M = 1.0$, $C_t = 1.0$, $C_i = 1.0$) 154

M16.2-9 Design Load Ratios for Exposed Timber Decks (Double and Single Tongue & Groove Decking) (Structural Calculations at Standard Reference Conditions: $C_D = 1.0$, $C_M = 1.0$, $C_t = 1.0$, $C_i = 1.0$) 155

M16.2-10 Design Load Ratios for Exposed Timber Decks (Butt-Joint Timber Decking) (Structural Calculations at Standard Reference Conditions: $C_D = 1.0$, $C_M = 1.0$, $C_t = 1.0$, $C_i = 1.0$) 155

LIST OF FIGURES

M5.1-1 Unbalanced and Balanced Layup Combinations 28

M5.2-1 Loading in the X-X and Y-Y Axes 30

M7.4-1 Design Span Determination 41

M7.4-2 Load Case Evaluations 43

M7.4-3 End Bearing Web Stiffeners (Bearing Block) 45

M7.4-4 Web Stiffener Bearing Interface 46

M7.4-5 Beveled End Cut 46

M7.4-6 Sloped Bearing Conditions (Low End) 47

M7.4-7 Sloped Bearing Conditions (High End) 48

M7.4-8 Lateral Support Requirements for Joists in Hangers 49

M7.4-9 Top Flange Hanger Support 49

M7.4-10 Connection Requirements for Face Nail Hangers 50

M7.4-11 Details for Vertical Load Transfer 51

M9.2-1 Structural Panel with Strength Direction Across Supports 60

M9.2-2 Example of Structural Panel in Bending.... 60

M9.2-3 Structural Panel with Axial Compression Load in the Plane of the Panel.... 64

M9.2-4 Through-the-Thickness Shear for Wood Structural Panels........ 64

M9.2-5 Planar (Rolling) Shear or Shear-in-the-Plane for Wood Structural Panels........ 64

M14.2-1 Shear Wall Drag Strut 94

M14.2-2 Shear Wall Special Case Drag Strut........ 95

M14.2-3 Diaphragm Drag Strut (Drag strut parallel to loads)........ 95

M14.2-4 Diaphragm Chord Forces 96

M14.3-1 Overturning Forces (no dead load) 97

M14.3-2 Overturning Forces (with dead load) 97

M16.1-1 One-Hour Fire-Resistive Wood Wall Assembly (WS4-1.1) (2x4 Wood Stud Wall - 100% Design Load - ASTM E119/NFPA 251) 105

M16.1-2 One-Hour Fire-Resistive Wood Wall Assembly (WS6-1.1) (2x6 Wood Stud Wall - 100% Design Load - ASTM E119/NFPA 251) 106

M16.1-3 One-Hour Fire-Resistive Wood Wall Assembly (WS6-1.2) (2x6 Wood Stud Wall - 100% Design Load - ASTM E119/NFPA 251) 107

M16.1-4 One-Hour Fire-Resistive Wood Wall Assembly (WS6-1.4) (2x6 Wood Stud Wall - 100% Design Load - ASTM E119/NFPA 251) 108

M16.1-5 One-Hour Fire-Resistive Wood Wall Assembly (WS4-1.2) (2x4 Wood Stud Wall - 100% Design Load - ASTM E119/NFPA 251 109

M16.1-6 One-Hour Fire-Resistive Wood Wall Assembly (WS4-1.3) (2x4 Wood Stud Wall - 78% Design Load - ASTM E119/NFPA 251) 110

M16.1-7 One-Hour Fire-Resistive Wood Wall Assembly (WS6-1.3) (2x6 Wood Stud Wall - 100% Design Load - ASTM E119/NFPA 251) 111

M16.1-8 One-Hour Fire-Resistive Wood Wall Assembly (WS6-1.5) (2x6 Wood Stud Wall - 100% Design Load - ASTM E119/NFPA 25) 112

M16.1-9 Two-Hour Fire-Resistive Wood Wall Assembly (WS6-2.1) (2x6 Wood Stud Wall - 100% Design Load - ASTM E119/NFPA 251) 113

M16.1-10 One-Hour Fire-Resistive Wood Floor/Ceiling Assembly (2x10 Wood Joists 16" o.c. – Gypsum Directly Applied or on Optional Resilient Channels) 114

M16.1-11 One-Hour Fire-Resistive Wood Floor/Ceiling Assembly (2x10 Wood Joists 16" o.c. – Suspended Acoustical Ceiling Panels) 115

M16.1-12 One-Hour Fire-Resistive Wood Floor/Ceiling Assembly (2x10 Wood Joists 16" o.c. – Gypsum on Resilient Channels) 116

M16.1-13 One-Hour Fire-Resistive Wood Floor/Ceiling Assembly (2x10 Wood Joists 24" o.c. – Gypsum on Resilient Channels) 117

M16.1-14 Two-Hour Fire-Resistive Wood Floor/Ceiling Assembly (2x10 Wood Joists 16" o.c. – Gypsum Directly Applied with Second Layer on Resilient Channels) 118

M16.1-15 One-Hour Fire-Resistive Ceiling Assembly (WIJ-1.1) (Floor/Ceiling - 100% Design Load - 1-Hour Rating - ASTM E119/NFPA 251) 124

M16.1-16 One-Hour Fire-Resistive Ceiling Assembly (WIJ-1.2) (Floor/Ceiling - 100% Design Load - 1 Hour Rating - ASTM E119/NFPA 251) 125

M16.1-17 One-Hour Fire-Resistive Ceiling Assembly (WIJ-1.3) (Floor/Ceiling - 100% Design Load - 1-Hour Rating - ASTM E119/NFPA 251) 126

M16.1-18 One-Hour Fire-Resistive Ceiling Assembly (WIJ-1.4) (Floor/Ceiling - 100% Design Load - 1-Hour Rating - ASTM E119/NFPA 251) 127

M16.1-19 One-Hour Fire-Resistive Ceiling Assembly (WIJ-1.5) (Floor/Ceiling - 100% Design Load - 1-Hour Rating - ASTM E119/NFPA 251) 128

M16.1-20 One-Hour Fire-Resistive Ceiling Assembly (WIJ-1.6) (Floor/Ceiling - 100% Design Load - 1-Hour Rating - ASTM E119/NFPA 251) 129

M16.1-21 Two-Hour Fire-Resistive Ceiling Assembly (WIJ-2.1) (Floor/Ceiling - 100% Design Load - 2-Hour Rating - ASTM E119/NFPA 251) 130

M16.1-22 Cross Sections of Possible One-Hour Area Separations 139

M16.1-23 Examples of Through-Penetration Firestop Systems 142

M16.3-1   Beam to Column Connection - Connection Not Exposed to Fire 159

M16.3-2   Beam to Column Connection - Connection Exposed to Fire Where Appearance is a Factor 159

M16.3-3   Ceiling Construction 159

M16.3-4   Beam to Column Connection - Connection Exposed to Fire Where Appearance is Not a Factor 159

M16.3-5   Column Connections Covered 160

M16.3-6   Beam to Girder - Concealed Connection 160

M1: GENERAL REQUIREMENTS FOR STRUCTURAL DESIGN

M1.1	Products Covered in This Manual	2
M1.2	General Requirements	2
	M1.2.1 Bracing	2
M1.3	Design Procedures	2

AMERICAN FOREST & PAPER ASSOCIATION

M1.1 Products Covered in This Manual

This Manual was developed with the intention of covering all structural applications of wood-based products and their connections that meet the requirements of the referenced standards. The Manual is a dual format document incorporating design provisions for both allowable stress design (ASD) and load and resistance factor design (LRFD). Design information is available for the following list of products. Each product chapter contains information for use with this Manual and the *National Design Specification® (NDS®) for Wood Construction*. Chapters are organized to parallel the chapter format of the *NDS*.

- Sawn Lumber   Chapter 4
- Structural Glued Laminated Timber   Chapter 5
- Round Timber Poles and Piles   Chapter 6
- Prefabricated Wood I-Joists   Chapter 7

- Structural Composite Lumber   Chapter
- Wood Structural Panels   Chapter
- Mechanical Connections   Chapter 1
- Dowel-Type Fasteners   Chapter 1
- Split Ring and Shear Plate Connectors   Chapter 1
- Timber Rivets   Chapter 1.
- Shear Walls and Diaphragms   Chapter 1

An additional Supplement, entitled *Special Desig Provisions for Wind and Seismic (SDPWS)*, has bee developed to cover materials, design, and construction o wood members, fasteners, and assemblies to resist win and seismic forces.

M1.2 General Requirements

This Manual is organized as a multi-part package for maximum flexibility for the design engineer. Included in this package are:

- *NDS* and Commentary; and, *NDS Supplement: Design Values for Wood Construction,*
- *Special Design Provisions for Wind and Seismic (SDPWS)* and Commentary,
- *Structural Wood Design Solved Example Problems.*

M1.2.1 Bracing

Design considerations related to both temporary an permanent bracing differ among product types. Specifi discussion of bracing is included in the product chapter.

M1.3 Design Procedures

The *NDS* is a dual format specification incorporating design provisions for ASD and LRFD. Behavioral equations, such as those for member and connection design, are the same for both ASD and LRFD. Adjustment factor tables include applicable factors for determining an adjusted ASD design value or an adjusted LRFD design value. NDS Appendix N – (Mandatory) Load and Resistance Factor Design (LRFD) outlines requirements that are unique to LRFD and adjustment factors for LRFD.

The basic design equations for ASD or LRFD require that the specified product reference design value meet or exceed the actual (applied) stress or other effect imposed by the specified loads. In ASD, the reference design values are set very low, and the nominal load magnitudes are set at once-in-a-lifetime service load levels. This combina tion produces designs that maintain high safety levels ye remain economically feasible.

From a user's standpoint, the design process is simi lar using LRFD. The most obvious difference betwee LRFD and ASD is that both the adjusted design value and load effect values in ASD will be numerically muc lower than in LRFD. The adjusted design values are lowe because they are reduced by significant internal safet adjustments. The load effects are lower because they ar nominal (service) load magnitudes. The load combinatio equations for use with ASD and LRFD are given in th model building codes.

M2: DESIGN VALUES FOR STRUCTURAL MEMBERS

2

M2.1	General Information	4
M2.2	Reference Design Values	4
M2.3	Adjustment of Design Values	4

M2.1 General Information

Structural wood products are provided to serve a wide range of end uses. Some products are marketed through commodity channels where the products meet specific standards and the selection of the appropriate product is the responsibility of the user.

Other products are custom manufactured to meet the specific needs of a given project. Products in this category are metal plate connected wood trusses and custom structural glued laminated timbers. Design of the individual members is based on criteria specified by the architect or engineer of record on the project. Manufacture of these products is performed in accordance with the product manufacturing standards. Engineering of these product normally only extends to the design of the product themselves. Construction-related issues, such as load path analysis and erection bracing, remain the responsibility of the professional of record for the project.

M2.2 Reference Design Values

Reference design value designates the allowable stress design value based on normal load duration. To avoid confusion, the descriptor "reference" is used and serves as a reminder that design value adjustment factors are applicable for design values in accordance with referenced conditions specified in the *NDS* – such as normal load duration.

Reference design values for sawn lumber and structural glued laminated timber are contained in the *NDS Supplement: Design Values for Wood Construction*. Reference design values for round timber poles and piles dowel-type fasteners, split ring and shear plate connectors and timber rivets are contained in the *NDS*. Reference design values for all other products are typically contained in the manufacturer's code evaluation report.

M2.3 Adjustment of Design Values

Adjusted design value designates reference design values which have been multiplied by adjustment factors. Basic requirements for design use terminology applicable to both ASD and LRFD. In equation format, this takes the standard form $f_b \leq F_b'$ which is applicable to either ASD or LRFD. Reference design values (F_b, F_t, F_v, F_c, $F_{c\perp}$, E, E_{min}) are multiplied by adjustment factors to determine adjusted design values (F_b', F_t', F_v', F_c', $F_{c\perp}'$, E', E_{min}').

Reference conditions have been defined such that a majority of wood products used in interior or in protected environments will require no adjustment for moisture, temperature, or treatment effects.

Moisture content (MC) reference conditions are 19% or less for sawn lumber products. The equivalent limit for glued products (structural glued laminated timber, structural composite lumber, prefabricated wood I-joists, and wood structural panels) is defined as 16% MC or less.

Temperature reference conditions include sustained temperatures up to 100°F. Note that it has been traditionally assumed that these reference conditions also include common building applications in desert locations where daytime temperatures will often exceed 100°F. Examples of applications that may exceed the reference temperature range include food processing or other industrial buildings.

Tabulated design values and capacities are for untreated members. Tabulated design values and capacities also apply to wood products pressure treated by an approved process and preservative except as specified for load duration factors.

An unincised reference condition is assumed. For members that are incised to increase penetration of preservative chemicals, use the incising adjustment factor given in the product chapter.

The effects of fire retardant chemical treatment on strength shall be accounted for in the design. Reference design values, including connection design values, for lumber and structural glued laminated timber pressure treated with fire retardant chemicals shall be obtained from the company providing the treatment and redrying service. The impact load duration factor shall not apply to structural members pressure-treated with fire retardant chemicals.

M3: DESIGN PROVISIONS AND EQUATIONS

3

M3.1   General   6

M3.2   Bending Members - General   6

M3.3   Bending Members - Flexure   6

M3.4   Bending Members - Shear   6

M3.5   Bending Members - Deflection   7

M3.6   Compression Members   7

M3.7   Solid Columns   8

M3.8   Tension Members   8

        M3.8.1   Tension Parallel to Grain   8

        M3.8.2   Tension Perpendicular to Grain   8

M3.9   Combined Bending and Axial Loading   9

M3.10   Design for Bearing   10

M3.1 General

This Chapter covers design of members for bending, compression, tension, combined bending and axial loads, and bearing.

M3.2 Bending Members – General

This section covers design of members stressed primarily in flexure (bending). Examples of such members include primary framing members (beams) and secondary framing members (purlins, joists). Products commonly used in these applications include glulam, solid sawn lumber, structural composite lumber, and prefabricated I-joists.

Bending members are designed so that no design capacity is exceeded under applied loads. Strength criteria for bending members include bending moment, shear, local buckling, lateral torsional buckling, and bearing.

See specific product chapters for moment and shear capacities (joist and beam selection tables) and reference bending and shear design values.

M3.3 Bending Members – Flexure

The basic equation for moment design of bending members is:

$$M' \geq M \qquad \text{(M3.3-1)}$$

where:

M' = adjusted moment capacity

M = bending moment

The equation for calculation of adjusted moment capacity is:

$$M' = F_b' \, S \qquad \text{(M3.3-2)}$$

where:

S = section modulus, in.^{3}

F_b' = adjusted bending design value, psi. See product chapters for applicable adjustment factors.

M3.4 Bending Members – Shear

The basic equation for shear design of bending members is:

$$V' \geq V \qquad \text{(M3.4-1)}$$

where:

V' = adjusted shear capacity parallel to grain, lbs

V = shear force, lbs

The equation for calculation of shear capacity is:

$$V' = F_v' \, Ib/Q \qquad \text{(M3.4-2)}$$

which, for rectangular unnotched bending members, reduces to:

$$V' = 2/3 \, (F_v') \, A \qquad \text{(M3.4-3)}$$

where:

I = moment of inertia, in.^{4}

A = area, in.^{2}

F_v' = adjusted shear design value, psi. See product chapters for applicable adjustment factors.

M3.5 Bending Members – Deflection

Users should note that design of bending members is often controlled by serviceability limitations rather than strength. Serviceability considerations, such as deflection and vibration, are often designated by the authority having jurisdiction.

For a simple span uniformly loaded rectangular member, the equation used for calculating mid-span deflection is:

$$\Delta = \frac{5w\ell^4}{384EI} \qquad (M3.5\text{-}1)$$

where:

Δ = deflection, in.

w = uniform load, lb/in.

ℓ = span, in.

EI = stiffness of beam section, lb-in.^{2}

Values of modulus of elasticity, E, and moment of inertia, I, for lumber and structural glued laminated timber for use in the preceding equation can be found in the *NDS Supplement*. Engineered wood products such as I-joists and structural composite lumber will have EI values published in individual manufacturer's product literature or evaluation reports. Some manufacturers might publish "true" E which would require additional computations to account for shear deflection. See *NDS* Appendix F for information on shear deflection. See product chapters for more details about deflection calculations.

M3.6 Compression Members

This section covers design of members stressed primarily in compression parallel to grain. Examples of such members include columns, truss members, and diaphragm chords.

Information in this section is limited to the case in which loads are applied concentrically to the member. Provisions of *NDS* 3.9 or *NDS* Chapter 15 should be used if loads are eccentric or if the compressive forces are applied in addition to bending forces.

The *NDS* differentiates between solid, built-up, and spaced columns. In this context built-up columns are assembled from multiple pieces of similar members connected in accordance with *NDS* 15.3.

A spaced column must comply with provisions of *NDS* 15.2. Note that this definition includes main column elements, spacer blocks with their connectors and end blocks with shear plate or split ring connectors.

Compression Parallel to Grain

The basic equation for design of compression members is:

$$P' \geq P \qquad (M3.6\text{-}1)$$

where:

P' = adjusted compression parallel to grain capacity, lbs

P = compressive force, lbs

The complete equation for calculation of adjusted compression capacity is:

$$P' = F_c'A \qquad (M3.6\text{-}2)$$

where:

A = area, in.^{2}

F_c' = adjusted compression parallel to grain design value, psi. See product chapters for applicable adjustment factors.

Special Considerations

Net Section Calculation

As in design of tension members, compression members should be checked both on a gross section and a net section basis (see *NDS* 3.6.3).

Bearing Capacity Checks

Design for bearing is addressed in *NDS* 3.10.

Radial Compression in Curved Members

Stresses induced in curved members under load include a component of stress in the direction of the radius of curvature. Radial compression is a specialized design consideration that is addressed in *NDS* 5.4.1.

M3.7 Solid Columns

Slenderness Considerations and Stability

The user is cautioned that stability calculations are highly dependent upon boundary conditions assumed in the analysis. For example, the common assumption of a pinned-pinned column is only accurate or conservative if the member is restrained against sidesway. If sidesway is possible and a pinned-free condition exists, the value of K_e in *NDS* 3.7.1.2 doubles (see *NDS* Appendix Table G1 for recommended buckling length coefficients, K_e) and the computed adjusted compression parallel to grain capacity decreases.

M3.8 Tension Members

This section covers design of members stressed primarily in tension parallel to grain. Examples of such members include shear wall end posts, truss members, and diaphragm chords.

The designer is advised that use of wood members in applications that induce tension perpendicular to grain stresses should be avoided.

M3.8.1 Tension Parallel to Grain

The basic equation for design of tension members is:

$$T' \geq T \qquad\qquad (M3.8\text{-}1)$$

where:

> T' = adjusted tension parallel to grain capacity, lbs
>
> T = tensile force, lbs

The equation for calculation of adjusted tension capacity is:

$$T' = F_t'A \qquad\qquad (M3.8\text{-}2)$$

where:

> A = area, in.^{2}
>
> F_t' = adjusted tension design value, psi. See product chapters for applicable adjustment factors.

Net Section Calculation

Design of tension members is often controlled by the ability to provide connections to develop tensile force within the member. In the area of connections, one must design not only the connection itself (described in detail in Chapter M10) but also the transfer of force across the net section of the member. One method for determining these stresses is provided in *NDS* Appendix E.

M3.8.2 Tension Perpendicular to Grain

Radial Stress in Curved Members

Stresses induced in curved members under load include a component of stress in the direction of the radius of curvature. This stress is traditionally called radial tension. Radial stress design is a specialized consideration that is covered in *NDS* 5.4.1 and is explained in detail in the American Institute of Timber Construction (AITC) *Timber Construction Manual*.

M3.9 Combined Bending and Axial Loading

This section covers design of members stressed under combined bending and axial loads. The applicable strength criteria for these members is explicit in the *NDS* equations - limiting the sum of various stress ratios to less than or equal to unity.

Bending and Axial Tension

For designs in which the axial load is in tension rather than compression, the designer should use *NDS* Equations 3.9-1 and 3.9-2.

Bending and Axial Compression

The equation for design of members under bending plus compression loads is given below in terms of load and moment ratios:

$$\left(\frac{P}{P'}\right)^2 + \frac{M_1}{M_1'\left(1-\frac{P}{P_{E1}}\right)} + \frac{M_2}{M_2'\left[1-\frac{P}{P_{E2}}-\left(\frac{M_1}{M_E}\right)^2\right]} \leq 1.0 \qquad \text{(M3.9-1)}$$

where

P′ = adjusted compression capacity determined per M3.6, lbs

P = compressive force determined per M3.6, lbs

M_1' = adjusted moment capacity (strong axis) determined per M3.3, in.-lbs

M_1 = bending moment (strong axis) determined per M3.3, in.-lbs

M_2' = adjusted moment capacity (weak axis) determined per M3.3, in.-lbs

M_2 = bending moment (weak axis) determined per M3.3, in.-lbs

P_{E1} = F_{cE1} A = critical column buckling capacity (strong axis) determined per *NDS* 3.9.2, lbs

P_{E2} = F_{cE2} A = critical column buckling capacity (weak axis) determined per *NDS* 3.9.2, lbs

M_E = F_{bE} S = critical beam buckling capacity determined per *NDS* 3.9.2, in.-lbs

Members must be designed by multiplying all applicable adjustment factors by the reference design values for the product. See M3.3 and M3.6 for discussion of applicable adjustment factors for bending or compression, respectively.

Design Techniques

A key to understanding design of members under combined bending and axial loads is that components of the design equation are simple ratios of compressive force (or moment) to compression capacity (or moment capacity). Note that the compression term in this equation is squared. This is the result of empirical test data. Moderate compressive forces do not have as large an impact on capacity (under combined loads) as previously thought. It is believed that this is the result of compressive "reinforcing" of what would otherwise be a tensile failure mode in bending.

M3.10 Design for Bearing

Columns often transfer large forces within a structural system. While satisfaction of column strength criteria is usually the primary concern, the designer should also check force transfer at the column bearing.

For cases in which the column is bearing on another wood member, especially if bearing is perpendicular to grain, this calculation will often control the design.

The basic equation for bearing design is:

$$R' \geq R \qquad (M3.10\text{-}1)$$

where:

R′ = adjusted compression perpendicular to grain capacity, lbs

R = compressive force or reaction, lbs

The equation for calculation of adjusted compression perpendicular to grain capacity is:

$$R' = F_{c\perp}'A \qquad (M3.10\text{-}2)$$

where:

A = area, in.^{2}

F_c' = adjusted compression perpendicular to grain design value, psi. See product chapters for applicable adjustment factors.

M4: SAWN LUMBER

4

M4.1	General	12
M4.2	Reference Design Values	12
M4.3	Adjustment of Reference Design Values	13
M4.4	Special Design Considerations	14
M4.5	Member Selection Tables	17
M4.6	Examples of Capacity Table Development	25

M4.1 General

Product Information

Structural lumber products are well-known throughout the construction industry. The economic advantages of lumber often dictate its choice as a preferred building material.

Lumber is available in a wide range of species, grades, sizes, and moisture contents. Structural lumber products are typically specified by either the stress level required or by the species, grade, and size required.

This Chapter provides information for designing structural lumber products in accordance with the *NDS*.

Common Uses

Structural lumber and timbers have been a primary construction material throughout the world for many centuries. They are the most widely used framing material for housing in North America.

In addition to use in housing, structural lumber finds broad use in commercial and industrial construction. Its high strength, universal availability, and cost saving attributes make it a viable option in most low- and mid-rise construction projects.

Structural lumber is used as beams, columns, headers, joists, rafters, studs, and plates in conventional construction. In addition to its use in lumber form, structural lumber is used to manufacture structural glued laminated beams, trusses, and wood I-joists.

Availability

Structural lumber is a widely available construction material. However, to efficiently specify structural lumber for individual construction projects, the specifier should be aware of the species, grades, and sizes available locally. The best source of this information is your local lumber supplier.

M4.2 Reference Design Values

General

The *NDS Supplement* provides reference design values for design of sawn lumber members. These design values are used when manual calculation of member capacity is required and must be used in conjunction with the adjustment factors specified in *NDS* 4.3.

Reference Design Values

Reference design values are provided in the *NDS Supplement* as follows:

NDS *Supplement* Table Number	
4A and 4B	Visually graded dimension lumber
4C	Mechanically graded dimension lumber
4D	Visually graded timbers
4E	Visually graded decking
4F	Non-North American visually graded dimension lumber

M4.3 Adjustment of Reference Design Values

To generate member design capacities, reference design values for sawn lumber are multiplied by adjustment factors and section properties per Chapter M3. Applicable adjustment factors for sawn lumber are defined in *NDS* 4.3. Table M4.3-1 shows the applicability of adjustment factors for sawn lumber in a slightly different format for the designer.

Table M4.3-1   Applicability of Adjustment Factors for Sawn Lumber

Allowable Stress Design	Load and Resistance Factor Design
$F_b' = F_b\, C_D\, C_M\, C_t\, C_L\, C_F\, C_{fu}\, C_i\, C_r$	$F_b' = F_b\, C_M\, C_t\, C_L\, C_F\, C_{fu}\, C_i\, C_r\, K_F\, \phi_b\, \lambda$
$F_t' = F_t\, C_D\, C_M\, C_t\, C_F\, C_i$	$F_t' = F_t\, C_M\, C_t\, C_F\, C_i\, K_F\, \phi_t\, \lambda$
$F_v' = F_v\, C_D\, C_M\, C_t\, C_i$	$F_v' = F_v\, C_M\, C_t\, C_i\, K_F\, \phi_v\, \lambda$
$F_{c\perp}' = F_{c\perp}\, C_M\, C_t\, C_i\, C_b$	$F_{c\perp}' = F_{c\perp}\, C_M\, C_t\, C_i\, C_b\, K_F\, \phi_c\, \lambda$
$F_c' = F_c\, C_D\, C_M\, C_t\, C_F\, C_i\, C_P$	$F_c' = F_c\, C_M\, C_t\, C_F\, C_i\, C_P\, K_F\, \phi_c\, \lambda$
$E' = E\, C_M\, C_t\, C_i$	$E' = E\, C_M\, C_t\, C_i$
$E_{min}' = E_{min}\, C_M\, C_t\, C_i\, C_T$	$E_{min}' = E_{min}\, C_M\, C_t\, C_i\, C_T\, K_F\, \phi_s$

Bending Member Example

For unincised, straight, laterally supported bending members stressed in edgewise bending in single member use and used in a normal building environment (meeting the reference conditions of *NDS* 2.3 and 4.3), the adjusted design values reduce to:

For ASD:

$$F_b' = F_b\, C_D\, C_F$$

$$F_v' = F_v\, C_D$$

$$F_{c\perp}' = F_{c\perp}\, C_b$$

$$E' = E$$

$$E_{min}' = E_{min}$$

For LRFD:

$$F_b' = F_b\, C_F\, K_F\, \phi_b\, \lambda$$

$$F_v' = F_v\, K_F\, \phi_v\, \lambda$$

$$F_{c\perp}' = F_{c\perp}\, C_b\, K_F\, \phi_c\, \lambda$$

$$E' = E$$

$$E_{min}' = E_{min}\, K_F\, \phi_s$$

Axially Loaded Member Example

For unincised axially loaded members used in a normal building environment (meeting the reference conditions of *NDS* 2.3 and 4.3) designed to resist tension or compression loads, the adjusted tension or compression design values reduce to:

For ASD:

$$F_c' = F_c\, C_D\, C_F\, C_P$$

$$F_t' = F_t\, C_D\, C_F$$

$$E_{min}' = E_{min}$$

For LRFD:

$$F_c' = F_c\, C_F\, C_P\, K_F\, \phi_c\, \lambda$$

$$F_t' = F_t\, C_F\, K_F\, \phi_t\, \lambda$$

$$E_{min}' = E_{min}\, K_F\, \phi_s$$

M4.4 Special Design Considerations

General

With proper detailing and protection, structural lumber can perform well in a variety of environments. One key to proper detailing is planning for the natural shrinkage and swelling of wood members as they are subjected to various drying and wetting cycles. While moisture changes have the largest impact on lumber dimensions, some designs must also check the effects of temperature on dimensions as well.

Dimensional Changes

Table M4.4-1 is extracted from more precise scientific and research reports on these topics. The coefficients are conservative (yielding more shrinkage and expansion than one might expect for most species). This level of information should be adequate for common structural applications. Equations are provided in this section for those designers who require more precise calculations.

Design of wood members and assemblies for fire resistance is discussed in Chapter M16.

Table M4.4-1   Approximate Moisture and Thermal Dimensional Changes

Description	Radial or Tangential Direction
Dimensional change due to moisture content change[1]	1% change in dimension per 4% change in MC
Dimensional change due to temperature change[2]	20×10^{-6} in./in. per degree F

1. Corresponding longitudinal direction shrinkage/expansion is about 1% to 5% of that in radial and tangential directions.
2. Corresponding longitudinal direction coefficient is about 1/10 as large as radial and tangential.

Equations for Computing Moisture and Thermal Shrinkage/Expansion

Due to Moisture Changes

For more precise computation of dimensional changes due to changes in moisture, the change in radial (R), tangential (T), and volumetric (V) dimensions due to changes in moisture content can be calculated as:

$$X = X_o \left(\Delta MC \right) e_{ME} \qquad \text{(M4.4-1)}$$

where:

X_0 = initial dimension or volume

X = new dimension or volume

ΔMC = moisture content change (%)

e_{ME} = coefficient of moisture expansion: linear (in./in./%MC) or volumetric (in.³/in.³/%MC)

and:

$$\Delta MC = M - M_o \qquad \text{(M4.4-2)}$$

where:

M_0 = initial moisture content % ($M_0 \leq FSP$)

M = new moisture content % ($M \leq FSP$)

FSP = fiber saturation point

Values for e_{ME} and FSP are shown in Table M4.4-2.

Due to Temperature Changes

For more precise calculation of dimensional change due to changes in temperature, the shrinkage/expansion of solid wood including lumber and timbers can be calculated as:

$$X = X_o \left(\Delta T \right) e_{TE} \qquad \text{(M4.4-3)}$$

where:

X_0 = reference dimension at T_0

X = computed dimension at T

T_0 = reference temperature (°F)

T = temperature at which the new dimension is calculated (°F)

e_{TE} = coefficient of thermal expansion (in./in./°F)

and:

$$\Delta T = T - T_o \qquad \text{(M4.4-4)}$$

where:

$-60°F \leq T_o \leq 130°F$

The coefficient of thermal expansion of ovendry wood parallel to grain ranges from about 1.7×10^{-6} to 2.5×10^{-6} per °F.

The linear expansion coefficients across the grain (radial and tangential) are proportional to wood density. These coefficients are about five to ten times greater than the parallel-to-the-grain coefficients and are given as:

Radial:

$$e_{TE} = \left[18(G) + 5.5\right]\left(10^{-6}\,\text{in./in./}^{\circ}F\right) \qquad \text{(M4.4-5)}$$

Tangential:

$$e_{TE} = \left[18(G) + 10.2\right]\left(10^{-6}\,\text{in./in./}^{\circ}F\right) \qquad \text{(M4.4-6)}$$

where:

G = tabulated specific gravity for the species.

Table M4.4-2   Coefficient of Moisture Expansion, e_{ME}, and Fiber Saturation Point, FSP, for Solid Woods

Species	e_{ME} Radial (in./in./%)	Tangential (in./in./%)	Volumetric (in.³/in.³/%)	FSP (%)
Alaska Cedar	0.0010	0.0021	0.0033	28
Douglas Fir-Larch	0.0018	0.0033	0.0050	28
Englemann Spruce	0.0013	0.0024	0.0037	30
Redwood	0.0012	0.0022	0.0032	22
Red Oak	0.0017	0.0038	0.0063	30
Southern Pine	0.0020	0.0030	0.0047	26
Western Hemlock	0.0015	0.0028	0.0044	28
Yellow Poplar	0.0015	0.0026	0.0041	31

Table M4.4-3 provides the numerical values for e_{TE} for the most commonly used commercial species or species groups.

Wood that contains moisture reacts to varying temperature differently than does dry wood. When moist wood is heated, it tends to expand because of normal thermal expansion and to shrink because of loss in moisture content. Unless the wood is very dry initially (perhaps 3% or 4% MC or less), the shrinkage due to moisture loss on heating will be greater than the thermal expansion, so the net dimensional change on heating will be negative. Wood at intermediate moisture levels (about 8% to 20%) will expand when first heated, then gradually shrink to a volume smaller than the initial volume, as the wood gradually loses water while in the heated condition.

Even in the longitudinal (grain) direction, where dimensional change due to moisture change is very small, such changes will still predominate over corresponding dimensional changes due to thermal expansion unless the wood is very dry initially. For wood at usual moisture levels, net dimensional changes will generally be negative after prolonged heating.

Calculation of actual changes in dimensions can be accomplished by determining the equilibrium moisture content of wood at the temperature value and relative humidity of interest. Then the relative dimensional changes due to temperature change alone and moisture content change alone are calculated. By combining these two changes the final dimension of lumber and timber can be established.

Table M4.4-3   Coefficient of Thermal Expansion, e_{TE}, for Solid Woods

Species	e_{TE} Radial (10^{-6} in./in./°F)	Tangential (10^{-6} in./in./°F)
California Redwood	13	18
Douglas Fir-Larch[1]	15	19
Douglas Fir, *South*	14	19
Eastern Spruce	13	18
Hem-Fir[1]	13	18
Red Oak	18	22
Southern Pine	15	20
Spruce-Pine-Fir	13	18
Yellow Poplar	14	18

1. Also applies when species name includes the designation "North."

Durability

Designing for durability is a key part of the architectural and engineering design of the building. This issue is particularly important in the design of buildings that use poles and piles. Many design conditions can be detailed to minimize the potential for decay; for other problem conditions, preservative-treated wood or naturally durable species should be specified.

This section does not cover the topic of designing for durability in detail. There are many excellent texts on the topic, including AF&PA's *Design of Wood Structures for Permanence, WCD No. 6*. Designers are advised to use this type of information to assist in designing for "difficult" design areas, such as:

- structures in moist or humid conditions
- where wood comes in contact with concrete or masonry
- where wood members are supported in steel hangers or connectors in which condensation could collect
- anywhere that wood is directly or indirectly exposed to the elements
- where wood, if it should ever become wet, could not naturally dry out.

This list is not intended to be all-inclusive – it is merely an attempt to alert designers to special conditions that may cause problems when durability is not considered in the design.

M4.5 Member Selection Tables

General

Member selection tables provide ASD capacities for many common designs. Before using the selection tables, refer to the footnotes to be certain that the tables are appropriate for the application.

The tables in this section provide design capacity values for structural lumber and timbers. Moment capacity, M', shear capacity, V', and bending stiffness, EI, are provided for strong-axis bending. Tension capacity, T', and compression capacity, P', are also tabulated. The applicable load duration factor, C_D, is indicated in each of the tables.

Footnotes are provided to allow conversion to load and resistance factor design (LRFD) capacities. See *NDS* Appendix N for more details on LRFD.

For manual calculation, two approaches are possible: 1) review the design equations in the chapter and modify the tabulated values as necessary; or 2) compute design capacity directly from the reference design values and adjustment factors.

To compute the design capacity for a specific condition, apply the design equations directly. Reference design values are provided in Chapter 4 of the *NDS Supplement* and design adjustment factors are provided in *NDS* 4.3.

ASD Capacity Tables

ASD capacity tables are provided as follows:

Table M4.5-1 = tension members
Table M4.5-2 = compression members (timbers)
Table M4.5-3 = bending members (lumber)
Table M4.5-4 = bending members (timber)

Refer to the selection table checklist to see whether your design condition meets the assumptions built into the tabulated values. Note that the load duration factor, C_D, is shown at the top of the table. Thus, the member design capacity can be used directly to select a member that meets the design requirement.

Examples of the development of the capacity table values are shown in M4.6.

Compression member tables are based on concentric axial loads only and pin-pin end conditions. Bending member tables are based on uniformly distributed loads on a simple span beam. Values from the compression or tension member tables and bending member tables cannot be combined in an interaction equation to determine combined bending and axial loads. See *NDS* 3.9 for more information.

Table M4.5-1a ASD Tension Member Capacity (T'), Structural Lumber[1,2]

2-inch nominal thickness Visually Graded Lumber (1.5 inch dry dressed size), $C_D = 1.0$.
4-inch nominal thickness Visually Graded Lumber (3.5 inch dry dressed size), $C_D = 1.0$.

Species	Width Nominal in.	Width Actual in.	Tension Member Capacity, T' (lbs) Visually Graded Lumber 2" nominal thickness Select Structural	No. 1	No. 2	No. 3	4" nominal thickness Select Structural	No. 1	No. 2
Douglas Fir-Larch	4	3.5	7,870	5,310	4,520	2,550	18,300	12,400	10,500
	6	5.5	10,700	7,230	6,160	3,480	25,000	16,800	14,300
	8	7.25	13,000	8,000	7,500	4,240	30,400	20,500	17,500
	10	9.25	15,260	10,300	8,770	4,960	35,600	24,000	20,400
	12	11.25	16,800	11,900	9,700	5,480	39,300	26,500	22,600
Hem-Fir	4	3.5	7,280	4,920	4,130	2,360	16,900	11,400	9,640
	6	5.5	9,900	6,700	5,630	3,210	23,100	15,600	13,100
	8	7.25	12,000	8,150	6,850	3,910	28,100	19,000	15,900
	10	9.25	14,100	9,530	8,010	4,570	32,900	22,200	18,600
	12	11.25	15,600	10,500	8,850	5,060	36,400	24,600	20,600
Southern Pine	4	3.5	8,400	5,510	4,330	2,490	19,600	12,800	10,100
	6	5.5	11,500	7,420	5,980	3,500	26,900	17,300	13,900
	8	7.25	14,100	8,970	7,060	4,350	32,900	20,900	16,400
	10	9.25	15,200	10,000	9,280	4,500	35,600	23,400	18,600
	12	11.25	17,700	11,300	9,280	5,480	41,300	26,500	21,600
Spruce-Pine-Fir	4	3.5	5,510	3,540	3,540	1,960	12,800	8,260	8,260
	6	5.5	7,510	4,820	4,280	2,680	17,500	11,200	11,200
	8	7.25	9,130	5,870	5,870	3,260	21,300	13,700	13,700
	10	9.25	10,600	6,860	6,860	3,810	24,900	16,000	16,000
	12	11.25	11,800	7,590	7,590	4,210	27,500	17,700	17,700

Species	Width Nominal in.	Width Actual in.	2" nominal thickness Construction	Standard	Utility	Stud	4" nominal thickness Construction	Standard	Stud
Douglas Fir-Larch	4	3.5	3,410	1,960	919	2,590	7,960	4,590	6,060
Hem-Fir	4	3.5	3,150	1,700	788	2,310	7,350	3,980	5,390
Southern Pine	4	3.5	3,280	1,830	919	2,490	7,560	4,280	5,810
Spruce-Pine-Fir	4	3.5	2,620	1,440	656	2,020	6,120	3,360	4,710

Table M4.5-1b ASD Tension Member Capacity (T'), Structural Lumber[1,2]

2-inch nominal thickness MSR Lumber (1.5 inch dry dressed size), $C_D = 1.0$.

Species	Width Nominal in.	Width Actual in.	Tension Member Capacity, T' (lbs) Machine Stress Rated Lumber 2" nominal thickness 1200f-1.2E	1350f-1.3E	1450f-1.3E	1650f-1.5E	2100f-1.8E	2250f-1.9E	2400f-2.0E
All Species	4	3.5	3,150	3,938	4,200	5,355	8,269	9,188	10,106
	6	5.5	4,950	6,188	6,600	8,415	12,994	14,438	15,881
	8	7.25	6,525	8,156	8,700	11,093	17,128	19,031	20,934
	10	9.25	8,325	10,406	11,100	14,153	21,853	24,281	26,709
	12	11.25	10,125	12,656	13,500	17,213	26,578	29,531	32,484

1. Multiply tabulated ASD capacity by 1.728 to obtain LRFD capacity ($\lambda = 0.8$). See *NDS* Appendix N for more information.
2. Tabulated values apply to members in a dry service condition, $C_M = 1.0$; normal temperature range, $C_t = 1.0$; and unincised members, $C_i = 1.0$.

Table M4.5-2a ASD Column Capacity[1,2,3,4,5] (P', P'$_x$, P'$_y$), Timbers

6-inch nominal thickness (5.5 inch dry dressed size), C$_D$ = 1.0.

Species	Column Length (ft)	Column Capacity (lbs)								
		Select Structural			No. 1			No. 2		
		6 x 6	6 x 8		6 x 6	6 x 8		6 x 6	6 x 8	
		6" width	8" width		6" width	8" width		6" width	8" width	
		(=5.5")	(=7.5")		(=5.5")	(=7.5")		(=5.5")	(=7.5")	
		P'	P'x	P'y	P'	P'x	P'y	P'	P'x	P'y
Douglas Fir-Larch	2	34,500	47,200	47,000	30,000	41,100	40,900	21,000	28,800	28,700
	4	33,400	46,400	45,500	29,200	40,500	39,800	20,500	28,400	28,000
	6	31,100	45,000	42,500	27,600	39,500	37,600	19,600	27,800	26,700
	8	27,300	42,700	37,300	24,800	37,800	33,800	18,000	26,800	24,500
	10	22,300	39,200	30,400	20,900	35,300	28,500	15,700	25,400	21,400
	12	17,500	34,600	23,900	16,800	31,800	22,900	13,000	23,400	17,700
	14	13,700	29,500	18,700	13,300	27,800	18,200	10,500	20,900	14,300
	16	10,900	24,700	14,800	10,700	23,700	14,600	8,500	18,200	11,500
Hem-Fir	2	29,200	40,000	39,800	25,500	34,900	34,800	17,300	23,600	23,600
	4	28,200	39,300	38,500	24,800	34,400	33,800	16,900	23,400	23,000
	6	26,200	38,100	35,800	23,300	33,500	31,800	16,100	22,900	22,000
	8	22,800	36,000	31,100	20,800	31,900	28,400	14,900	22,100	20,300
	10	18,400	32,900	25,100	17,400	29,700	23,700	13,100	21,000	17,800
	12	14,300	28,800	19,600	13,800	26,600	18,900	10,900	19,400	14,800
	14	11,200	24,300	15,200	10,900	23,000	14,900	8,800	17,400	12,000
	16	8,800	20,200	12,000	8,700	19,500	11,900	7,200	15,200	9,800
Southern Pine	2	28,500	39,000	38,900	24,800	33,900	33,800	15,800	21,600	21,500
	4	27,700	38,500	37,800	24,200	33,500	33,000	15,500	21,400	21,100
	6	26,200	37,500	35,700	23,100	32,800	31,500	15,000	21,000	20,400
	8	23,500	35,900	32,100	21,200	31,600	28,900	14,100	20,500	19,200
	10	19,900	33,500	27,100	18,400	29,900	25,100	12,700	19,700	17,400
	12	16,000	30,200	21,800	15,200	27,500	20,700	11,000	18,500	15,000
	14	12,700	26,400	17,300	12,300	24,500	16,700	9,200	17,100	12,500
	16	10,100	22,400	13,800	9,900	21,300	13,500	7,600	15,300	10,300
Spruce-Pine-Fir	2	24,000	32,900	32,700	21,000	28,800	28,700	15,000	20,600	20,500
	4	23,400	32,400	31,900	20,500	28,400	28,000	14,700	20,300	20,100
	6	22,100	31,600	30,100	19,600	27,800	26,700	14,100	19,900	19,300
	8	19,900	30,300	27,100	18,000	26,800	24,500	13,100	19,300	17,900
	10	16,800	28,300	23,000	15,700	25,400	21,400	11,600	18,400	15,800
	12	13,600	25,600	18,500	13,000	23,400	17,700	9,800	17,100	13,400
	14	10,800	22,400	14,700	10,500	20,900	14,300	8,000	15,500	11,000
	16	8,600	19,100	11,800	8,500	18,200	11,500	6,500	13,700	8,900

P'$_x$ values are based on a column continuously braced against weak axis buckling.
P'$_y$ values are based on a column continuously braced against strong axis buckling.
To obtain LRFD capacity, see *NDS* Appendix N.
Tabulated values apply to members in a dry service condition, C$_M$ = 1.0; normal temperature range, C$_t$ = 1.0; and unincised members, C$_i$ = 1.0.
Column capacities are based on concentric axial loads only and pin-pin end conditions (K$_e$ = 1.0 per *NDS* Appendix Table G1).

Table M4.5-2b ASD Column Capacity (P', P'ₓ, P'ᵧ), Timbers[1,2,3,4,5]

8-inch nominal thickness (7.5 inch dry dressed size), $C_D = 1.0$.

Species	Column Length (ft)	Select Structural 8 x 8 8" width (=7.5") P'	Select Structural 8 x 10 10" width (=9.5") P'x	Select Structural 8 x 10 10" width (=9.5") P'y	No. 1 8 x 8 8" width (=7.5") P'	No. 1 8 x 10 10" width (=9.5") P'x	No. 1 8 x 10 10" width (=9.5") P'y	No. 2 8 x 8 8" width (=7.5") P'	No. 2 8 x 10 10" width (=9.5") P'x	No. 2 8 x 10 10" width (=9.5") P'y
Douglas Fir-Larch	2	64,400	81,700	81,500	56,000	71,100	70,900	39,200	49,800	49,700
	4	63,300	80,900	80,200	55,200	70,500	70,000	38,800	49,400	49,100
	6	61,400	79,500	77,800	53,800	69,400	68,200	37,900	48,800	48,000
	8	58,300	77,300	73,800	51,500	67,800	65,300	36,600	47,800	46,300
	10	53,500	74,000	67,800	48,100	65,400	60,900	34,600	46,500	43,800
	12	47,200	69,600	59,800	43,400	62,200	55,000	31,900	44,600	40,400
	14	40,200	63,800	50,900	37,900	58,000	48,000	28,500	42,100	36,100
	16	33,600	57,000	42,600	32,300	52,800	40,900	24,800	39,100	31,400
Hem-Fir	2	54,600	69,200	69,100	47,600	60,400	60,300	32,200	40,900	40,800
	4	53,600	68,500	67,900	46,900	59,900	59,400	31,900	40,600	40,400
	6	51,900	67,300	65,800	45,600	58,900	57,800	31,200	40,100	39,500
	8	49,100	65,300	62,200	43,600	57,500	55,200	30,100	39,300	38,200
	10	44,800	62,400	56,800	40,400	55,300	51,200	28,600	38,300	36,200
	12	39,200	58,400	49,700	36,200	52,400	45,900	26,400	36,800	33,500
	14	33,200	53,200	42,000	31,400	48,600	39,700	23,700	34,900	30,100
	16	27,600	47,300	34,900	26,500	44,000	33,600	20,800	32,500	26,300
Southern Pine	2	53,200	67,500	67,400	46,200	58,600	58,600	29,400	37,300	37,300
	4	52,500	66,900	66,500	45,700	58,200	57,900	29,200	37,100	36,900
	6	51,100	65,900	64,700	44,700	57,500	56,600	28,700	36,800	36,300
	8	48,900	64,400	62,000	43,100	56,300	54,600	27,900	36,200	35,400
	10	45,700	62,200	57,800	40,700	54,700	51,600	26,800	35,400	34,000
	12	41,200	59,100	52,200	37,500	52,500	47,500	25,300	34,400	32,000
	14	35,900	55,000	45,500	33,400	49,600	42,400	23,300	33,000	29,500
	16	30,600	50,200	38,800	29,100	46,000	36,800	20,900	31,300	26,500
Spruce-Pine-Fir	2	44,800	56,800	56,800	39,200	49,800	49,700	28,000	35,500	35,500
	4	44,200	56,400	56,000	38,800	49,400	49,100	27,700	35,300	35,100
	6	43,100	55,500	54,600	37,900	48,800	48,000	27,200	34,900	34,400
	8	41,300	54,300	52,300	36,600	47,800	46,300	26,300	34,300	33,400
	10	38,600	52,400	48,900	34,600	46,500	43,800	25,100	33,400	31,800
	12	34,900	49,900	44,200	31,900	44,600	40,400	23,400	32,300	29,600
	14	30,500	46,500	38,600	28,500	42,100	36,100	21,200	30,700	26,800
	16	26,000	42,500	33,000	24,800	39,100	31,400	18,700	28,800	23,700

1. P'ₓ values are based on a column continuously braced against weak axis buckling.
2. P'ᵧ values are based on a column continuously braced against strong axis buckling.
3. To obtain LRFD capacity, see *NDS* Appendix N.
4. Tabulated values apply to members in a dry service condition, $C_M = 1.0$; normal temperature range, $C_t = 1.0$; and unincised members, $C_i = 1.0$.
5. Column capacities are based on concentric axial loads only and pin-pin end conditions ($K_e = 1.0$ per *NDS* Appendix Table G1).

Table M4.5-2c ASD Column Capacity (P', P'$_x$, P'$_y$), Timbers[1,2,3,4,5]

10-inch nominal thickness (9.5 inch dry dressed size), $C_D = 1.0$.

Species	Column Length (ft)	Select Structural 10 x 10 10" width (=9.5") P'	Select Structural 10 x 12 12 width (=11.5") P'x	Select Structural 10 x 12 12 width (=11.5") P'y	No. 1 10 x 10 10" width (=9.5") P'	No. 1 10 x 12 12 width (=11.5") P'x	No. 1 10 x 12 12 width (=11.5") P'y	No. 2 10 x 10 10" width (=9.5") P'	No. 2 10 x 12 12 width (=11.5") P'x	No. 2 10 x 12 12 width (=11.5") P'y
Douglas Fir-Larch	2	103,500	125,400	125,200	90,000	109,000	109,000	63,000	76,400	76,300
	4	102,500	124,600	124,000	89,300	108,400	108,000	62,600	76,000	75,700
	6	100,700	123,100	121,900	87,900	107,400	106,400	61,800	75,300	74,800
	8	97,900	121,000	118,500	85,900	105,800	103,900	60,600	74,400	73,300
	10	93,800	117,900	113,500	82,900	103,500	100,300	58,800	73,100	71,200
	12	88,100	113,800	106,700	78,800	100,500	95,400	56,500	71,300	68,400
	14	80,800	108,300	97,800	73,400	96,600	88,900	53,400	69,100	64,600
	16	72,200	101,500	87,500	66,900	91,600	81,000	49,500	66,200	59,900
Hem-Fir	2	87,700	106,300	106,200	76,500	92,700	92,600	51,800	62,700	62,700
	4	86,800	105,600	105,100	75,800	92,100	91,800	51,400	62,400	62,200
	6	85,200	104,300	103,100	74,600	91,200	90,300	50,800	61,900	61,500
	8	82,700	102,400	100,100	72,800	89,700	88,100	49,800	61,200	60,300
	10	79,000	99,600	95,700	70,100	87,700	84,800	48,500	60,100	58,700
	12	73,900	95,900	89,500	66,400	85,000	80,400	46,600	58,800	56,400
	14	67,400	91,000	81,600	61,500	81,500	74,500	44,200	57,000	53,500
	16	59,900	84,900	72,500	55,700	77,000	67,500	41,100	54,700	49,800
Southern Pine	2	85,500	103,600	103,500	74,300	90,000	89,900	47,300	57,300	57,200
	4	84,800	103,000	102,600	73,700	89,500	89,300	47,000	57,100	56,900
	6	83,500	102,000	101,100	72,800	88,800	88,100	46,600	56,700	56,400
	8	81,600	100,500	98,700	71,400	87,700	86,400	45,800	56,100	55,500
	10	78,700	98,300	95,300	69,300	86,100	83,900	44,900	55,400	54,300
	12	74,800	95,500	90,600	66,500	84,000	80,500	43,500	54,400	52,700
	14	69,700	91,700	84,400	62,800	81,300	76,000	41,800	53,100	50,600
	16	63,500	87,000	76,900	58,200	77,900	70,500	39,600	51,500	47,900
Spruce-Pine-Fir	2	72,000	87,200	87,200	63,000	76,400	76,300	45,000	54,500	54,500
	4	71,400	86,800	86,400	62,600	76,000	75,700	44,700	54,300	54,200
	6	70,400	85,900	85,200	61,800	75,300	74,800	44,200	53,900	53,500
	8	68,700	84,600	83,200	60,600	74,400	73,300	43,400	53,300	52,600
	10	66,400	82,900	80,400	58,800	73,100	71,200	42,400	52,400	51,300
	12	63,200	80,500	76,500	56,500	71,300	68,400	40,900	51,300	49,500
	14	59,000	77,400	71,400	53,400	69,100	64,600	38,900	49,900	47,100
	16	53,800	73,500	65,200	49,500	66,200	59,900	36,500	48,100	44,100

P'$_x$ values are based on a column continuously braced against weak axis buckling.
P'$_y$ values are based on a column continuously braced against strong axis buckling.
To obtain LRFD capacity, see *NDS* Appendix N.
Tabulated values apply to members in a dry service condition, $C_M = 1.0$; normal temperature range, $C_t = 1.0$; and unincised members, $C_i = 1.0$.
Column capacities are based on concentric axial loads only and pin-pin end conditions ($K_e = 1.0$ per *NDS* Appendix Table G1).

Table M4.5–3a ASD Bending Member Capacity (M', C$_r$M', V', and EI), Structural Lumber[1,2]

2-inch nominal thickness (1.5 inch dry dressed size), C$_D$ = 1.0, C$_L$ = 1.0.

| | Size (b x d) | | Select Structural | | | | No. 2 | | | |
| | Nominal | Actual | M' (Single) | Cr M' (Repetitive) | V' | x 10⁶ EI | M' | Cr M' (Repetitive) | V' | x 10⁶ EI |
Species	(in.)	(in.)	lb-in.	lb-in.	lbs	lb-in.²	lb-in.	lb-in.	lbs	lb-in.²
Douglas Fir-Larch	2 x 4	1.5 x 3.5	6,890	7,920	630	10	4,130	4,750	630	9
	2 x 6	1.5 x 5.5	14,700	17,000	990	40	8,850	10,200	990	33
	2 x 8	1.5 x 7.25	23,700	27,200	1,310	91	14,200	16,300	1,310	76
	2 x 10	1.5 x 9.25	35,300	40,600	1,670	188	21,200	24,400	1,670	158
	2 x 12	1.5 x 11.25	47,500	54,600	2,030	338	28,500	32,700	2,030	285
Hem-Fir	2 x 4	1.5 x 3.5	6,430	7,400	525	9	3,900	4,490	525	7
	2 x 6	1.5 x 5.5	13,800	15,900	825	33	8,360	9,610	825	27
	2 x 8	1.5 x 7.25	22,100	25,400	1,090	76	13,400	15,400	1,090	62
	2 x 10	1.5 x 9.25	32,900	37,900	1,390	158	20,000	23,000	1,390	129
	2 x 12	1.5 x 11.25	44,300	50,900	1,690	285	26,900	30,900	1,690	231
Southern Pine	2 x 4	1.5 x 3.5	8,730	10,000	613	10	4,590	5,280	613	9
	2 x 6	1.5 x 5.5	19,300	22,200	963	37	9,450	10,900	963	33
	2 x 8	1.5 x 7.25	30,200	34,800	1,270	86	15,800	18,100	1,270	76
	2 x 10	1.5 x 9.25	43,900	50,400	1,620	178	22,500	25,800	1,620	158
	2 x 12	1.5 x 11.25	60,100	69,100	1,970	320	30,800	35,500	1,970	285
Spruce-Pine-Fir	2 x 4	1.5 x 3.5	5,740	6,600	473	8	4,020	4,620	473	8
	2 x 6	1.5 x 5.5	12,300	14,100	743	31	8,600	9,890	743	29
	2 x 8	1.5 x 7.25	19,700	22,700	979	71	13,800	15,900	979	67
	2 x 10	1.5 x 9.25	29,400	33,800	1,250	148	20,600	23,700	1,250	139
	2 x 12	1.5 x 11.25	39,600	45,500	1,520	267	27,700	31,800	1,520	249

Table M4.5–3b ASD Bending Member Capacity (M', C$_r$M', V', and EI), Structural Lumber[1,2]

4-inch nominal thickness (3.5 inch dry dressed size), C$_D$ = 1.0, C$_L$ = 1.0.

| | Size (b x d) | | Select Structural | | | | No. 2 | | | |
| | Nominal | Actual | M' (Single) | Cr M' (Repetitive) | V' | x 10⁶ EI | M' | Cr M' (Repetitive) | V' | x 10⁶ EI |
Species	(in.)	(in.)	lb-in.	lb-in.	lbs	lb-in.²	lb-in.	lb-in.	lbs	lb-in.²
Douglas Fir-Larch	4 x 4	3.5 x 3.5	16,100	18,500	1,470	24	9,650	11,100	1,470	20
	4 x 6	3.5 x 5.5	34,400	39,600	2,310	92	20,600	23,700	2,310	78
	4 x 8	3.5 x 7.25	59,800	68,800	3,050	211	35,900	41,300	3,050	178
	4 x 10	3.5 x 9.25	89,800	103,000	3,890	439	53,900	62,000	3,890	370
	4 x 12	3.5 x 11.25	122,000	140,000	4,730	789	73,100	84,100	4,730	664
Hem-Fir	4 x 4	3.5 x 3.5	15,000	17,300	1,230	20	9,110	10,500	1,230	16
	4 x 6	3.5 x 5.5	32,100	36,900	1,930	78	19,500	22,400	1,930	63
	4 x 8	3.5 x 7.25	55,800	64,200	2,540	178	33,900	39,000	2,540	144
	4 x 10	3.5 x 9.25	83,900	96,400	3,240	369	50,900	58,500	3,240	300
	4 x 12	3.5 x 11.25	114,000	131,000	3,940	664	69,000	79,400	3,940	540
Southern Pine	4 x 4	3.5 x 3.5	20,400	23,400	1,430	23	10,700	12,300	1,430	20
	4 x 6	3.5 x 5.5	45,000	51,700	2,250	87	22,100	25,400	2,250	78
	4 x 8	3.5 x 7.25	77,600	89,200	2,960	200	40,500	46,600	2,960	178
	4 x 10	3.5 x 9.25	113,000	129,000	3,780	416	57,600	66,300	3,780	369
	4 x 12	3.5 x 11.25	154,000	177,000	4,590	748	79,200	91,100	4,590	664
Spruce-Pine-Fir	4 x 4	3.5 x 3.5	13,400	15,400	1,100	19	9,380	10,800	1,100	18
	4 x 6	3.5 x 5.5	28,700	33,000	1,730	73	20,100	23,100	1,730	68
	4 x 8	3.5 x 7.25	49,800	57,300	2,280	167	34,900	40,100	2,280	156
	4 x 10	3.5 x 9.25	74,900	86,100	2,910	346	52,400	60,300	2,910	323
	4 x 12	3.5 x 11.25	102,000	117,000	3,540	623	71,100	81,700	3,540	581

1. Multiply tabulated M', C$_r$M', and V' capacity by 1.728 to obtain LRFD capacity (λ = 0.8). Tabulated EI is applicable for both ASD and LRFD. See *NDS* Append N for more information.
2. Tabulated values apply to members in a dry service condition, C$_M$ = 1.0; normal temperature range, C$_t$ = 1.0; and unincised members, C$_i$ = 1.0; members brac against buckling, C$_L$ = 1.0.

Table M4.5-4a ASD Bending Member Capacity (M', V', and EI), Timbers[1,2]

6-inch nominal thickness (5.5 inch dry dressed size), $C_D = 1.0$, $C_L = 1.0$.

Species	Size (b x d) Nominal (in.)	Actual (in.)	Select Structural M' (Single) lb-in.	V' lbs	x 10⁶ EI lb-in.²	No. 2 M' lb-in.	V' lbs	x 10⁶ EI lb-in.²
Douglas Fir-Larch	6 x 6	5.5 x 5.5	41,600	3,430	122	20,800	3,430	99
	6 x 8	5.5 x 7.5	77,300	4,680	309	38,700	4,680	251
	6 x 10	5.5 x 9.5	132,000	5,920	629	72,400	5,920	511
	6 x 12	5.5 x 11.5	194,000	7,170	1,120	106,000	7,170	906
	6 x 14	5.5 x 13.5	264,000	8,420	1,800	144,000	8,420	1,470
	6 x 16	5.5 x 15.5	342,000	9,660	2,730	187,000	9,660	2,220
Hem-Fir	6 x 6	5.5 x 5.5	33,300	2,820	99	15,900	2,820	84
	6 x 8	5.5 x 7.5	61,900	3,850	251	29,600	3,850	213
	6 x 10	5.5 x 9.5	10,800	4,880	511	55,800	4,880	432
	6 x 12	5.5 x 11.5	15,800	5,900	906	81,800	5,900	767
	6 x 14	5.5 x 13.5	214,000	6,930	1,470	111,000	6,930	1,240
	6 x 16	5.5 x 15.5	278,000	7,960	2,220	144,000	7,960	1,880
Southern Pine	6 x 6	5.5 x 5.5	41,600	3,330	114	23,600	3,330	92
	6 x 8	5.5 x 7.5	77,300	4,540	290	43,800	4,540	232
	6 x 10	5.5 x 9.5	124,000	5,750	589	70,300	5,750	472
Spruce-Pine-Fir	6 x 6	5.5 x 5.5	29,100	2,520	99	13,900	2,520	76
	6 x 8	5.5 x 7.5	54,100	3,440	251	25,800	3,440	193
	6 x 10	5.5 x 9.5	91,000	4,350	511	49,600	4,350	393

Table M4.5-4b ASD Bending Member Capacity (M', V', and EI), Timbers[1,2]

8-inch nominal thickness (7.5 inch dry dressed size), $C_D = 1.0$, $C_L = 1.0$.

Species	Size (b x d) Nominal (in.)	Actual (in.)	Select Structural M' (Single) lb-in.	V' lbs	x 10⁶ EI lb-in.²	No. 2 M' lb-in.	V' lbs	x 10⁶ EI lb-in.²
Douglas Fir-Larch	8 x 8	7.5 x 7.5	105,000	6,380	422	52,700	6,380	343
	8 x 10	7.5 x 9.5	169,000	8,080	857	84,600	8,080	697
	8 x 12	7.5 x 11.5	265,000	9,780	1,520	145,000	9,780	1,240
	8 x 14	7.5 x 13.5	360,000	11,500	2,460	197,000	11,500	2,000
	8 x 16	7.5 x 15.5	467,000	13,200	3,720	255,000	13,200	3,030
Hem-Fir	8 x 8	7.5 x 7.5	84,400	5,250	343	40,400	5,250	290
	8 x 10	7.5 x 9.5	135,000	6,650	697	64,900	6,650	589
	8 x 12	7.5 x 11.5	215,000	8,050	1,240	112,000	8,050	1,050
	8 x 14	7.5 x 13.5	292,000	9,450	2,000	152,000	9,450	1,690
	8 x 16	7.5 x 15.5	379,000	10,900	3,030	197,000	10,900	2,560
Southern Pine	8 x 8	7.5 x 7.5	105,000	6,190	396	59,800	6,190	316
	8 x 10	7.5 x 9.5	169,000	7,840	804	95,900	7,840	643
Spruce-Pine-Fir	8 x 8	7.5 x 7.5	73,800	4,690	343	35,200	4,690	264
	8 x 10	7.5 x 9.5	118,000	5,940	697	56,400	5,940	536

Multiply tabulated M' and V' capacity by 1.728 to obtain LRFD capacity ($\lambda = 0.8$). Tabulated EI is applicable for both ASD and LRFD. See *NDS* Appendix N for more information.

Tabulated values apply to members in a dry service condition, $C_M = 1.0$; normal temperature range, $C_t = 1.0$; and unincised members, $C_i = 1.0$; members braced against buckling, $C_L = 1.0$.

Table M4.5-4c ASD Bending Member Capacity (M', V', and EI), Timbers[1,2]

10-inch nominal thickness (9.5 inch dry dressed size), C_D = 1.0, C_L = 1.0.

Species	Size (b x d) Nominal (in.)	Size (b x d) Actual (in.)	Select Structural M' (Single) lb-in.	Select Structural V' lbs	Select Structural x 10⁶ EI lb-in.²	No. 2 M' lb-in.	No. 2 V' lbs	No. 2 x 10⁶ EI lb-in.²
	10 x 10	9.5 x 9.5	214,000	10,200	1,090	107,000	10,200	882
	10 x 12	9.5 x 11.5	314,000	12,400	1,930	157,000	12,400	1,570
Douglas Fir-Larch	10 x 14	9.5 x 13.5	456,000	14,500	3,120	249,000	14,500	2,530
	10 x 16	9.5 x 15.5	592,000	16,700	4,720	324,000	16,700	3,830
	10 x 18	9.5 x 17.5	744,000	18,800	6,790	407,000	18,800	5,520
	10 x 20	9.5 x 19.5	913,000	21,000	9,390	499,000	21,000	7,630
	10 x 10	9.5 x 9.5	171,000	8,420	882	82,200	8,420	747
	10 x 12	9.5 x 11.5	251,000	10,200	1,570	120,000	10,200	1,320
Hem-Fir	10 x 14	9.5 x 13.5	370,000	12,000	2,530	192,000	12,000	2,140
	10 x 16	9.5 x 15.5	418,000	13,700	3,830	250,000	13,700	3,240
	10 x 18	9.5 x 17.5	604,000	15,500	5,520	314,000	15,500	4,670
	10 x 20	9.5 x 19.5	742,000	17,300	7,630	385,000	17,300	6,460
	10 x 10	9.5 x 9.5	214,000	9,930	1,020	121,000	9,930	815
Southern Pine	10 x 12	9.5 x 11.5	314,000	12,000	1,810	178,000	12,000	1,440
	10 x 14	9.5 x 13.5	427,000	14,100	2,920	242,000	14,100	2,340
	10 x 10	9.5 x 9.5	150,000	7,520	882	71,000	7,520	679
Spruce-Pine-Fir	10 x 12	9.5 x 11.5	220,000	9,100	1,570	105,000	9,100	1,200
	10 x 14	9.5 x 13.5	313,000	10,700	2,530	171,000	10,700	1,950

Table M4.5-4d ASD Bending Member Capacity (M', V', and EI), Timbers[1,2]

Nominal dimensions > 10 inch (actual = nominal – 1/2 inch), C_D = 1.0, C_L = 1.0.

Species	Size (b x d) Nominal (in.)	Size (b x d) Actual (in.)	Select Structural M' (Single) lb-in.	Select Structural V' lbs	Select Structural x 10⁶ EI lb-in.²	No. 2 M' lb-in.	No. 2 V' lbs	No. 2 x 10⁶ EI lb-in.²
	12 x 12	11.5 x 11.5	380,000	15,000	2,330	190,000	15,000	1,890
	14 x 14	13.5 x 13.5	607,000	20,700	4,430	304,000	20,700	3,600
Douglas Fir-Larch	16 x 16	15.5 x 15.5	905,000	27,200	7,700	452,000	27,200	6,250
	18 x 18	17.5 x 17.5	1,280,000	34,700	12,500	642,000	34,700	10,200
	20 x 20	19.5 x 19.5	1,760,000	43,100	19,300	878,000	43,100	15,700
	12 x 12	11.5 x 11.5	304,000	12,300	1,890	146,000	12,300	1,600
	14 x 14	13.5 x 13.5	486,000	17,000	3,600	233,000	17,000	3,040
Hem-Fir	16 x 16	15.5 x 15.5	724,000	22,400	6,250	347,000	22,400	5,290
	18 x 18	17.5 x 17.5	1,030,000	28,600	10,200	493,000	28,600	8,600
	20 x 20	19.5 x 19.5	1,410,000	35,500	15,700	673,000	35,500	13,300
Southern Pine	12 x 12	11.5 x 11.5	380,000	14,500	2,190	215,000	14,500	1,750
	14 x 14	13.5 x 13.5	607,000	20,000	4,150	344,000	20,000	3,320
Spruce-Pine-Fir	12 x 12	11.5 x 11.5	266,000	11,000	1,890	127,000	11,000	1,460
	14 x 14	13.5 x 13.5	425,000	15,200	3,600	202,000	15,200	2,770

1. Multiply tabulated M' and V' capacity by 1.728 to obtain LRFD capacity (λ = 0.8). Tabulated EI is applicable for both ASD and LRFD. See *NDS* Appendix N f[...] more information.
2. Tabulated values apply to members in a dry service condition, C_M = 1.0; normal temperature range, C_t = 1.0; and unincised members, C_i = 1.0; members brac[...] against buckling, C_L = 1.0.

M4.6 Examples of Capacity Table Development

Tension Capacity Tables

The general design equation for tension members is:

$$T' \geq T$$

where:

    T = tension force due to design loads

    T′ = allowable tension capacity

Example M4.6-1: Application – tension member

  Species: Hem-Fir

    Size: 2 x 6 (1.5 in. by 5.5 in.)

  Grade: 1650f-1.5E MSR

    F_t: 1,020 psi

    A: 8.25 in.^{2}

Tension Capacity

$$T' = F_t'A$$
$$= (1,020)(8.25)$$
$$= 8,415 \text{ lbs}$$

Column Capacity Tables

The general design equation is:

$$P' \geq P$$

where:

    P = compressive force due to design loads

    P′ = allowable compression capacity

Axial Capacity

$$P' = C_p A F^*_c$$

where:

$$C_P = \frac{1 + \left(F_{cE} / F^*_c\right)}{2c} - \sqrt{\left(\frac{1 + \left(F_{cE} / F^*_c\right)}{2c}\right)^2 - \frac{F_{cE} / F^*_c}{c}}$$

$$F_{cE} = \frac{0.822E'}{\left(\ell_e / d\right)^2}$$

and:

F^*_c = reference compression design value multiplied by all applicable adjustment factors except C_p

  A = area

  C_P = column stability factor

  F_c' = adjusted parallel-to-grain compression design value

  E_{min}' = adjusted modulus of elasticity for column stability calculations

  c = 0.8 for solid sawn members

Example M4.6-2: Application – simple column

  Species: Douglas Fir-Larch

    Size: 6 x 8 (5.5 in. by 7.5 in.) by 12 ft.

  Grade: No. 1 (dry) Posts and Timbers

    F^*_c: 1,000 psi

    E_{min}': 580,000 psi

    A: 41.25 in^2

Column Capacity – x-axis

$$F_{cE} = \frac{0.822E_{min}'}{\left(\ell_e / d_x\right)^2}$$
$$= \frac{0.822(580,000)}{(144/7.5)^2}$$
$$= 1,293 \text{ psi}$$

$$C_{Px} = \frac{1 + (1,293/1,000)}{2(0.8)}$$
$$- \sqrt{\left(\frac{1 + (1,293/1,000)}{2(0.8)}\right)^2 - \frac{1,293/1,000}{0.8}}$$
$$= 0.772$$

$$P_x' = (0.772)(41.25)(1,000)$$
$$= 31,845 \text{ lb.}$$

Column Capacity – y-axis

$$F_{cE} = \frac{0.822 E_{min}'}{\left(\ell_e / d_y\right)^2}$$

$$= \frac{0.822(580,000)}{(144/5.5)^2}$$

$$= 696 \text{ psi}$$

$$C_{Py} = \frac{1+(696/1,000)}{2(0.8)}$$

$$-\sqrt{\left(\frac{1+(696/1,000)}{2(0.8)}\right)^2 - \frac{696/1,000}{0.8}}$$

$$= 0.556$$

$$P_x' = (0.556)(41.25)(1,000)$$

$$= 22,952 \text{ lb}$$

Bending Member Capacity Tables

The general design equation for flexural bending is:

$$M' \geq M$$

where:

  M = moment due to design loads
  M′ = allowable moment capacity

The general design equation for flexural shear is:

$$V' \geq V$$

where:

  V = shear force due to design loads
  V′ = allowable shear capacity

Example M4.6-3:   Application – structural lumber

Species: Douglas Fir-Larch
  Size: 2 x 6 (1.5 in. by 5.5 in.)
  Grade: No. 2
    C_F: 1.3 (size factor)
C_r:1.15 (repetitive member factor)
    F_b: 900 psi
    F_v: 180 psi
    E: 1,600,000 psi
    A: 8.25 in.^{2}
    S: 7.56 in.^{3}
    I: 20.80 in.^{4}

Moment Capacity

$$C_r M' = F_b C_F C_r S$$

$$= (900)(1.3)(1.15)(7.56)$$

$$= 10,172 \text{ lb - in.}$$

Shear Capacity

$$V' = F_v' A \left(\frac{2}{3}\right)$$

$$= (180)(8.25)\left(\frac{2}{3}\right)$$

$$= 990 \text{ lb}$$

Flexural Stiffness

$$EI = (1,600,000)(20.80) = 33 \times 10^6 \text{ lb - in.}^2$$

M5: STRUCTURAL GLUED LAMINATED TIMBER

5

M5.1	General	28
M5.2	Reference Design Values	30
M5.3	Adjustment of Reference Design Values	31
M5.4	Special Design Considerations	32

M5.1 General

Products Description

Structural glued laminated timber (glulam) is a structural member glued up from suitably selected and prepared pieces of wood either in a straight or curved form with the grain of all of the pieces parallel to the longitudinal axis of the member. The reference design values given in the *NDS Supplement* are applicable only to structural glued laminated timber members produced in accordance with *American National Standard for Wood Products — Structural Glued Laminated Timber*, ANSI/AITC A190.1.

Structural glued laminated timber members are produced in laminating plants by gluing together dry lumber, normally of 2-inch or 1-inch nominal thickness, under controlled temperature and pressure conditions. Members with a wide variety of sizes, profiles, and lengths can be produced having superior characteristics of strength, serviceability, and appearance. Structural glued laminated timber beams are manufactured with the strongest laminations on the bottom and top of the beam, where the greatest tension and compression stresses occur in bending. This allows a more efficient use of the lumber resource by placing higher grade lumber in zones that have higher stresses and lumber with less structural quality in lower stressed zones.

Structural glued laminated timber members are manufactured from several softwood species, primarily Douglas fir-larch, southern pine, hem-fir, spruce-pine-fir, eastern spruce, western woods, Alaska cedar, Durango pine, and California redwood. In addition, several hardwood species, including red oak, red maple, and yellow poplar, are also used. Standard structural glued laminated timber sizes are given in the *NDS Supplement*. Any length, up to the maximum length permitted by transportation and handling restrictions, is available.

A structural glued laminated timber member can be manufactured using a single grade or multiple grades of lumber, depending on intended use. In addition, a mixed-species structural glued laminated timber member is also possible. When the member is intended to be primarily loaded either axially or in bending with the loads acting parallel to the wide faces of the laminations, a single grade combination is recommended. On the other hand, a multiple grade combination provides better cost-effectiveness when the member is primarily loaded in bending due to loads applied perpendicular to the wide faces of the laminations.

On a multiple grade combination, a structural glued laminated timber member can be produced as either a balanced or unbalanced combination, depending on the geometrical arrangement of the laminations about the mid-depth of the member. As shown in Figure M5.1-1, a balanced combination is symmetrical about the mid-depth so both faces have the same reference bending design value. Unbalanced combinations are asymmetrical and when used as a beam, the face with a lower allowable bending stress is stamped as TOP. The balanced combination is intended for use in continuous or cantilevered over supports to provide equal capacity in both positive and negative bending. Whereas the unbalanced combination is primarily for use in simple span applications, they can also be used for short cantilever applications (cantilever less than 20% of the back span) or for continuous span applications when the design is controlled by shear or deflection.

Figure M5.1-1 Unbalanced and Balanced Layup Combinations

No. 2D	Tension Lam
No. 2	No. 1
No. 2	No. 2
No. 3	No. 3
No. 3	No. 3
No. 3	No. 3
No. 2	No. 2
No. 1	No. 1
Tension Lam	Tension Lam
Unbalanced	Balanced

Structural glued laminated timber members can be used as primary or secondary load-carrying components in structures. Table M5.1-1 lists economical spans for selected timber framing systems using structural glued laminated timber members in buildings. Other common uses of structural glued laminated timber members are for utility structures, pedestrian bridges, highway bridges, railroad bridges, marine structures, noise barriers, and towers. Table M5.1-1 may be used for preliminary design purposes to determine the economical span ranges for the selected framing systems. However, all systems require a more extensive analysis for final design.

Table M5.1-1	Economical Spans for Structural Glued Laminated Timber Framing Systems

Type of Framing System	Economical Spans (ft)
ROOF	
Simple Span Beams	
Straight or slightly cambered	10 to 100
Tapered, double tapered-pitched, or curved	25 to 105
Cantilevered Beams (Main span)	up to 90
Continuous Beams (Interior spans)	10 to 50
Girders	40 to 100
Three-Hinged Arches	
Gothic	40 to 100
Tudor	40 to 140
A-Frame	20 to 100
Three-centered, Parabolic, or Radial	40 to 250
Two-Hinged Arches	
Radial or Parabolic	50 to 200
Trusses (Four or more ply chords)	
Flat or parallel chord	50 to 150
Triangular or pitched	50 to 150
Bowstring (Continuous chord)	50 to 200
Trusses (Two or three ply chords)	
Flat or parallel chord	20 to 75
Triangular or pitched	20 to 75
Tied arches	50 to 200
Dome structures	200 to 500+
FLOOR	
Simple Span Beams	10 to 40
Continuous Beams (Individual spans)	10 to 40
HEADERS	
Windows and Doors	< 10
Garage Doors	9 to 18

Appearance Classifications

Structural glued laminated timber members are typically produced in four appearance classifications: Premium, Architectural, Industrial, and Framing. Premium and Architectural beams are higher in appearance qualities and are surfaced for a smooth finish ready for staining or painting. Industrial classification beams are normally used in concealed applications or in construction where appearance is not important. Framing classification beams are typically used for headers and other concealed applications in residential construction. Design values for structural glued laminated timber members are independent of the appearance classifications.

For more information and detailed descriptions of these appearance classifications and their typical uses, refer to APA EWS Technical Note Y110 or AITC Standard 110.

Availability

Structural glued laminated timber members are available in both custom and stock sizes. Custom beams are manufactured to the specifications of a specific project, while stock beams are made in common dimensions, shipped to distribution yards, and cut to length when the beam is ordered. Stock beams are available in virtually every major metropolitan area. Although structural glued laminated timber members can be custom fabricated to provide a nearly infinite variety of forms and sizes, the best economy is generally realized by using standard-size members as noted in the *NDS Supplement*. When in doubt, the designer is advised to check with the structural glued laminated timber supplier or manufacturer concerning the availability of a specific size prior to specification.

M5.2 Reference Design Values

Reference design values of structural glued laminated timber are affected by the layup of members composed of various grades of lumber as well as the direction of applied bending forces. As a result, different design values are assigned for structural glued laminated timber used primarily in bending (*NDS Supplement* Table 5A) and primarily in axial loading (*NDS Supplement* Table 5B). The reference design values are used in conjunction with the dimensions provided in Table 1C (western species) and Table 1D (southern pine) of the *NDS Supplement*, but are applicable to any size of structural glued laminated timber when the appropriate modification factors discussed in M5.3 are applied.

Reference design values are given in *NDS Supplement* Table 5A for bending about the X-X axis (see Figure M5.2-1). Although permitted, axial loading or bending about the Y-Y axis (also see Figure M5.2-1) is not efficient in using the structural glued laminated timber combinations given in *NDS Supplement* Table 5A. In such cases, the designer should select structural glued laminated timber from *NDS Supplement* Table 5B. Similarly, structural glued laminated timber combinations in *NDS Supplement* Table 5B are inefficiently utilized if the primary use is bending about the X-X axis.

The reference design values given in *NDS Supplement* Tables 5A and 5B are based on use under normal duration of load (10 years) and dry conditions (less than 16% moisture content). When used under other conditions, see *NDS* Chapter 5 for adjustment factors. The reference bending design values are based on members loaded as simple beams. When structural glued laminated timber is used in continuous or cantilevered beams, the reference bending design values given in *NDS Supplement* Table 5A for compression zone stressed in tension should be used for the design of stress reversal.

Figure M5.2-1　Loading in the X-X and Y-Y Axes

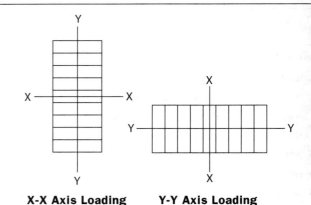

X-X Axis Loading　　　Y-Y Axis Loading

M5.3 Adjustment of Reference Design Values

The adjustment factors provided in the *NDS* are for non-reference end-use conditions and material modification effects. These factors shall be used to modify the reference design values when one or more of the specific end uses or material modification conditions are beyond the limits of the reference conditions given in the *NDS*.

Adjustment factors unique to structural glued laminated timber include the volume factor, C_V, and the curvature factor, C_c. Both are defined in Chapter 5 of the *NDS*.

To generate member design capacities, reference design values for structural glued laminated timber are multiplied by adjustment factors and section properties per Chapter M3. Applicable adjustment factors for structural glued laminated timber are defined in *NDS* 5.3. Table M5.3-1 shows the applicability of adjustment factors for structural glued laminated timber in a slightly different format for the designer.

Table M5.3-1   Applicability of Adjustment Factors for Structural Glued Laminated Timber[1]

Allowable Stress Design	Load and Resistance Factor Design
$F_b' = F_b\,C_D\,C_M\,C_t\,C_L\,C_V\,C_{fu}\,C_c$	$F_b' = F_b\,C_M\,C_t\,C_L\,C_V\,C_{fu}\,C_c\,K_F\,\phi_b\,\lambda$
$F_t' = F_t\,C_D\,C_M\,C_t$	$F_t' = F_t\,C_M\,C_t\,K_F\,\phi_t\,\lambda$
$F_v' = F_v\,C_D\,C_M\,C_t$	$F_v' = F_v\,C_M\,C_t\,K_F\,\phi_v\,\lambda$
$F_{c\perp}' = F_{c\perp}\,C_M\,C_t\,C_b$	$F_{c\perp}' = F_{c\perp}\,C_M\,C_t\,C_b\,K_F\,\phi_c\,\lambda$
$F_c' = F_c\,C_D\,C_M\,C_t\,C_P$	$F_c' = F_c\,C_M\,C_t\,C_P\,K_F\,\phi_c\,\lambda$
$E' = E\,C_M\,C_t$	$E' = E\,C_M\,C_t$
$E_{min}' = E_{min}\,C_M\,C_t$	$E_{min}' = E_{min}\,C_M\,C_t\,K_F\,\phi_s$

1. The beam stability factor, C_L, shall not apply simultaneously with the volume factor, C_V, for structural glued laminated timber bending members (see *NDS* 5.3.6). Therefore, the lesser of these adjustment factors shall apply.

Bending Member Example

For straight, laterally supported bending members loaded perpendicular to the wide face of the laminations and used in a normal building environment (meeting the reference conditions of *NDS* 2.3 and 5.3), the adjusted design values reduce to:

For ASD:

$$F_b' = F_b\,C_D\,C_V$$

$$F_v' = F_v\,C_D$$

$$F_{c\perp}' = F_{c\perp}\,C_b$$

$$E' = E$$

$$E_{min}' = E_{min}$$

For LRFD:

$$F_b' = F_b\,C_V\,K_F\,\phi_b\,\lambda$$

$$F_v' = F_v\,K_F\,\phi_v\,\lambda$$

$$F_{c\perp}' = F_{c\perp}\,C_b\,K_F\,\phi_c\,\lambda$$

$$E' = E$$

$$E_{min}' = E_{min}\,K_F\,\phi_s$$

Axially Loaded Member Example

For axially loaded members used in a normal building environment (meeting the reference conditions of *NDS* 2.3 and 5.3) designed to resist tension or compression loads, the adjusted tension or compression design values reduce to:

For ASD:

$$F_c' = F_c\,C_D\,C_P$$

$$F_t' = F_t\,C_D$$

$$E_{min}' = E_{min}$$

For LRFD:

$$F_c' = F_c\,C_P\,K_F\,\phi_c\,\lambda$$

$$F_t' = F_t\,K_F\,\phi_t\,\lambda$$

$$E_{min}' = E_{min}\,K_F\,\phi_s$$

5

M5: STRUCTURAL GLUED LAMINATED TIMBER

M5.4 Special Design Considerations

General

The section contains information concerning physical properties of structural glued laminated timber members including specific gravity and response to moisture or temperature.

In addition to designing to accommodate dimensional changes and detailing for durability, another significant issue in the planning of wood structures is that of fire performance, which is discussed in Chapter M16.

Specific Gravity

Table M5.4-1 provides specific gravity values for some of the most common wood species used for structural glued laminated timber. These values are used in determining various physical and connection properties. Further, weight factors are provided at four moisture contents. When the cross-sectional area (in.^{2}) is multiplied by the appropriate weight factor, it provides the weight of the structural glued laminated timber member per linear foot of length. For other moisture contents, the tabulated weight factors can be interpolated or extrapolated.

Structural glued laminated timber members often are manufactured using different species at different portions of the cross section. In this case the weight of the structural glued laminated timber may be computed by the sum of the products of the cross-sectional area and the weight factor for each species.

Dimensional Changes

See M4.4 for information on calculating dimensional changes due to moisture or temperature.

Durability

See M4.4 for information on detailing for durability.

Table M5.4–1 Average Specific Gravity and Weight Factor

Species Combination	Specific Gravity[1]	Weight Factor[2]			
		12%	15%	19%	25%
California Redwood (close grain)	0.44	0.195	0.198	0.202	0.208
Douglas Fir-Larch	0.50	0.235	0.238	0.242	0.248
Douglas Fir (South)	0.46	0.221	0.225	0.229	0.235
Eastern Spruce	0.41	0.191	0.194	0.198	0.203
Hem-Fir	0.43	0.195	0.198	0.202	0.208
Red Maple	0.58	0.261	0.264	0.268	0.274
Red Oak	0.67	0.307	0.310	0.314	0.319
Southern Pine	0.55	0.252	0.255	0.259	0.265
Spruce-Pine-Fir (North)	0.42	0.195	0.198	0.202	0.208
Yellow Poplar	0.43	0.213	0.216	0.220	0.226

1. Specific gravity is based on weight and volume when ovendry.
2. Weight factor shall be multiplied by net cross-sectional area in in.^{2} to obtain weight in pounds per lineal foot.

M6: ROUND TIMBER POLES AND PILES

6

M6.1	General	34
M6.2	Reference Design Values	34
M6.3	Adjustment of Reference Design Values	35
M6.4	Special Design Considerations	36

M6.1 General

Product Information

Timber **poles** are used extensively in post-frame construction and are also used architecturally. This Chapter is not for use with poles used in the support of utility lines. Timber **piles** are generally used as part of foundation systems.

Timber poles and piles offer many advantages relative to competing materials. As with other wood products, timber poles and piles offer the unique advantage of being the only major construction material that is a renewable resource.

Common Uses

Timber poles are used extensively in post-frame construction and are also used architecturally. This Chapter is not for use with poles used in the support of utility lines. Timber piles are generally used as part of foundation systems.

Timber poles and piles offer many advantages relative to competing materials. As with other wood product, timber poles and piles offer the unique advantage of bein the only major construction material that is a renewabl resource.

Availability

Timber piles are typically available in four specie: Pacific Coast Douglas-fir, southern pine, red oak, and re pine. However, local pile suppliers should be contacte because availability is dependent upon geographic loca tion.

Timber poles are supplied to the utility industry i a variety of grades and species. Because these poles ar graded according to ANSI 05.1, *Specifications and Dimer sions for Wood Poles*, they must be regraded according t ASTM D3200 if they are to be used with the *NDS*.

M6.2 Reference Design Values

General

The tables in *NDS* Chapter 6 provide reference design values for timber pole and pile members. These reference design values are used when manual calculation of member strength is required and shall be used in conjunction with adjustment factors specified in *NDS* Chapter 6.

Pole Reference Design Values

Reference design values for poles are provided in *NDS* Table 6B. These values, with the exception of F_c, are applicable for all locations in the pole. The F_c values are for the tip of the pole and can be increased for Pacific Coast Douglas-fir and southern pine poles in accordance with *NDS* 6.3.9.

Reference design values are applicable for wet exposure and for poles treated with a steam conditioning or Boultonizing process. For poles that are not treated, or are air-dried or kiln-dried prior to treating, the factors in *NDS* 6.3.5 shall be applied.

Pile Reference Design Values

Reference design values for piles are provided i *NDS* Table 6A. These values, with the exception of F are applicable at any location along the length of the pil The tabulated F_c values for Pacific Coast Douglas-fir an southern pine may be increased for locations other tha the tip as provided by *NDS* 6.3.9.

Reference design values are applicable for wet expo sures. These tabulated values are given for air-dried pile treated with a preservative using a steam conditioning c Boultonizing process. For piles that are not treated, or ar air-dried or kiln-dried prior to treating, the factors in *ND* 6.3.5 shall be applied.

To generate member design capacities, reference design values are multiplied by adjustment factors and section properties. Adjustment factors unique to round timber poles and piles include the untreated factor, C_u, the critical section factor, C_{cs}, and the single pile factor, C_{sp}. All are defined in Chapter 6 of the *NDS*.

To generate member design capacities, reference design values for round timber poles and piles are multiplied by adjustment factors and section properties per Chapter 13. Applicable adjustment factors for round timber poles and piles are defined in *NDS* 6.3. Table M6.3-1 shows the applicability of adjustment factors for round timber poles and piles in a slightly different format for the designer.

Table M6.3-1   Applicability of Adjustment Factors for Round Timber Poles and Piles[1]

Allowable Stress Design	Load and Resistance Factor Design
$F_c' = F_c\, C_D\, C_t\, C_u\, C_P\, C_{cs}\, C_{sp}$	$F_c' = F_c\, C_t\, C_u\, C_P\, C_{cs}\, C_{sp}\, K_F\, \phi_c\, \lambda$
$F_b' = F_b\, C_D\, C_t\, C_u\, C_F\, C_{sp}$	$F_t' = F_t\, C_t\, C_u\, C_F\, C_{sp}\, K_F\, \phi_b\, \lambda$
$F_v' = F_v\, C_D\, C_t\, C_u$	$F_v' = F_v\, C_t\, C_u\, K_F\, \phi_v\, \lambda$
$F_{c\perp}' = F_{c\perp}\, C_D^{1}\, C_t\, C_u\, C_b$	$F_{c\perp}' = F_{c\perp}\, C_t\, C_u\, C_b\, K_F\, \phi_c\, \lambda$
$E' = E\, C_t$	$E' = E\, C_t$
$E_{min}' = E_{min}\, C_t$	$E_{min}' = E_{min}\, C_t\, K_F\, \phi_s$

1. The C_D factor shall not apply to compression perpendicular to grain for poles.

...ially Loaded Pole or Pile Example

For single, axially loaded, treated poles or piles, fully laterally supported in two orthogonal directions, used in a normal environment (meeting the reference conditions of *NDS* 2.3 and 6.3), designed to resist compression loads only, and less than 13.5" in diameter, the adjusted compression design values reduce to:

For ASD:

$$F_c' = F_c\, C_D\, C_{sp}$$

For LRFD:

$$F_c' = F_c\, C_{sp}\, K_F\, \phi_c\, \lambda$$

M6.4 Special Design Considerations

With proper detailing and protection, timber poles and piles can perform well in a variety of environments. One key to proper detailing is planning for the natural shrinkage and swelling of wood products as they are subjected to various drying and wetting cycles. While moisture changes have the largest impact on product dimensions, some designs must also check the effects of temperature. See M4.4 for design information on dimensional changes due to moisture and temperature.

Durability issues related to piles are generally both more critical and more easily accommodated. Since piles are in constant ground contact, they cannot be "insulated" from contact with moisture – thus, the standard reference condition for piles is preservatively treated. The importance of proper treatment processing of piles cannot be overemphasized. See M4.4 for more information about durability.

In addition to designing to accommodate dimensional changes and detailing for durability, another significant issue in the planning of wood structures is that of fire performance, which is discussed in Chapter M16.

M7: PREFABRICATED WOOD I-JOISTS

M7.1   General   38

M7.2   Reference Design Values   38

M7.3   Adjustment of Reference
        Design Values   40

M7.4   Special Design Considerations   41

7

M7.1 General

Product Information

I-joists are exceptionally stiff, lightweight, and capable of long spans. Holes may be easily cut in the web according to manufacturer's recommendations, allowing ducts and utilities to be run through the joist. I-joists are dimensionally stable and uniform in size, with no crown. This keeps floors quieter, reduces field modifications, and eliminates rejects in the field. I-joists may be field cut to proper length using conventional methods and tools.

Manufacturing of I-joists utilizes the geometry of the cross section and high strength components to maximize the strength and stiffness of the wood fiber. Flanges are manufactured from solid sawn lumber or structural composite lumber, while webs typically consist of plywood or oriented strand board. The efficient utilization of raw materials, along with high-quality exterior adhesives and state of the art quality control procedures, result in an extremely consistent product that maximizes environmental benefits as well.

Wood I-joists are produced as proprietary products which are covered by code acceptance reports by one or all of the model building codes. Acceptance reports and product literature should be consulted for current design information.

Common Uses

Prefabricated wood I-joists are widely used as a frame ing material for housing in North America. I-joists are made in different grades and with various processes and can be utilized in various applications. Proper design required to optimize performance and economics.

In addition to use in housing, I-joists find increasing use in commercial and industrial construction. The high strength, stiffness, wide availability, and cost saving at tributes make them a viable alternative in most low-rise construction projects.

Prefabricated wood I-joists are typically used as floor and roof joists in conventional construction. In addition I-joists are used as studs where long lengths and high strengths are required.

Availability

To efficiently specify I-joists for individual constru tion projects, consideration should be given to the size an the required strength of the I-joist. Sizes vary with each individual product. The best source of this information your local lumber supplier, distribution center, or I-joi manufacturer. Proper design is facilitated through the u of manufacturer's literature and specification softwa available from I-joist manufacturers.

M7.2 Reference Design Values

Introduction to Design Values

As stated in *NDS* 7.2, each wood I-joist manufacturer develops its own proprietary design values. The derivation of these values is reviewed by the applicable building code authority. Since materials, manufacturing processes, and product evaluations may differ between the various manufacturers, selected design values are only appropriate for the specific product and application.

To generate the design capacity of a given product, the manufacturer of that product evaluates test data. The design capacity is then determined per ASTM D5055.

The latest model building code agency evaluation reports are a reliable source for wood I-joist design values. These reports list accepted design values for shear, moment, stiffness, and reaction capacity based on minimum bearing. In addition, evaluation reports note the limitations on web holes, concentrated loads, and requirements for web stiffeners.

Bearing/Reaction Design

Tabulated design capacities reflect standard condition and must be modified as discussed in *NDS* Chapter 7 obtain adjusted capacity values.

Bearing lengths at supports often control the desig capacity of an I-joist. Typically minimum bearing length are used to establish design parameters. In some case additional bearing is available and can be verified in installation. Increased bearing length means that the joi can support additional loading, up to the value limited the shear capacity of the web material and web joint. Bo interior and exterior reactions must be evaluated.

Use of web stiffeners may be required and typical increases the bearing capacity of the joist. Correct i stallation is required to obtain the specified capacitie Additional loading from walls above will load the joist bearing, further limiting the capacity of the joist if prop nd detailing is not followed. Additional information on earing specifics can be found in M7.4.

Adjusted bearing capacities, R_r', are determined in the ame empirical fashion as is allowable shear.

Shear Design

At end bearing locations, critical shear is the vertical shear at the ends of the design span. The practice of neglecting all uniform loads within a distance from the end support equal to the joist depth, commonly used for other wood materials, is not applicable to end supports for wood I-joists. At locations of continuity, such as interior supports of multi-span I-joists, the critical shear location for several wood I-joist types is located a distance equal to the depth of the joist from the centerline of bearing (uniform loads only). A cantilevered portion of a wood I-joist is generally not considered a location of continuity (unless the cantilever length exceeds the joist depth) and vertical shear at the cantilever bearing is the critical shear. Individual manufacturers, or the appropriate evaluation reports, should be consulted for reference to shear design at locations of continuity.

Often, the adjusted shear capacities, V_r', are based on other considerations such as bottom flange bearing length or the installation of web stiffeners or bearing blocks.

Moment Design

Adjusted moment capacities, M_r', of I-joists are determined from empirical testing of a completely assembled joist or by engineering analysis supplemented by tension testing the flange component. If the flange contains end jointed material, the allowable tension value is the lesser of the joint capacity or the material capacity.

Because flanges of a wood I-joist can be highly stressed, field notching of the flanges is not allowed. Similarly, excessive nailing or the use of improper nail sizes can cause flange splitting that will also reduce capacity. The manufacturer should be contacted when evaluating a damaged flange.

Deflection Design

Wood I-joists, due to their optimized web materials, are susceptible to the effects of shear deflection. This component of deflection can account for as much as 15% to 30% of the total deflection. For this reason, both bending and shear deflection are considered in deflection design. A typical deflection calculation for simple span wood I-joists under uniform load is shown in Equation M7.2-1.

Deflection = Bending Component + Shear Component

$$\Delta = \frac{5w\ell^4}{384EI} + \frac{w\ell^2}{k} \qquad (M7.2\text{-}1)$$

Individual manufacturers provide equations in a similar format. Values for use in the preceding equation can be found in the individual manufacturer's evaluation reports. For other load and span conditions, an approximate answer can be found by using conventional bending deflection equations adjusted as follows:

$$\text{Deflection} = \text{Bending Deflection} \left(1 + \frac{384\,EI}{5\ell^2 k}\right)$$

where:

w = uniform load in pounds per lineal inch

ℓ = design span, in.

EI = joist moment of inertia times flange modulus of elasticity

k = shear deflection coefficient

Since wood I-joists can have long spans, the model building code maximum live load deflection criteria may not be appropriate for many floor applications. Many wood I-joist manufacturers recommend using stiffer criteria, such as L/480 for residential floor construction and L/600 for public access commercial applications such as office floors. The minimum code required criteria for storage floors and roof applications is normally adequate.

M7.3 Adjustment of Reference Design Values

General

Member design capacity is the product of reference design values and adjustment factors. Reference design values for I-joists are discussed in M7.2.

The design values listed in the evaluation reports are generally applicable to dry use conditions. Less typical conditions, such as high moisture, high temperatures, or pressure impregnated chemical treatments, typically resul in strength and stiffness adjustments different from thos used for sawn lumber. *NDS* 7.3 outlines adjustments t design values for I-joists; however, individual wood I-jois manufacturers should be consulted to verify appropriat adjustments. Table M7.3-1 shows the applicability o adjustment factors for prefabricated wood I-joists in slightly different format for the designer.

Table M7.3-1 Applicability of Adjustment Factors for Prefabricated Wood I-Joists

Allowable Stress Design	Load and Resistance Factor Design
$M_r' = M_r\,C_D\,C_M\,C_t\,C_L\,C_r$	$M_r' = M_r\,C_M\,C_t\,C_L\,C_r\,K_F\,\phi_b\,\lambda$
$V_r' = V_r\,C_D\,C_M\,C_t$	$V_r' = V_r\,C_D\,C_M\,C_t\,K_F\,\phi_v\,\lambda$
$R_r' = R_r\,C_D\,C_M\,C_t$	$R_r' = R_r\,C_M\,C_t\,K_F\,\phi_v\,\lambda$
$EI' = EI\,C_M\,C_t$	$EI' = EI\,C_M\,C_t$
$EI_{min}' = EI_{min}\,C_M\,C_t$	$EI_{min}' = EI_{min}\,C_M\,C_t\,K_F\,\phi_s$

Bending Member Example

For fully laterally supported bending members loaded in strong axis bending and used in a normal building environment (meeting the reference conditions of *NDS* 2.3 and 7.3), the adjusted design values reduce to:

For ASD:

$$M_r' = M_r\,C_D$$
$$V_r' = V_r\,C_D$$
$$R_r' = R_r\,C_D$$
$$EI' = EI$$
$$K' = K$$

For LRFD:

$$M_r' = M_r\,K_F\,\phi_b\,\lambda$$
$$V_r' = V_r\,K_F\,\phi_v\,\lambda$$
$$R_r' = R_r\,K_F\,\phi_v\,\lambda$$
$$EI' = EI$$
$$K' = K$$

The user is cautioned that manufacturers may no permit the use of some applications and/or treatment: Unauthorized treatments can void a manufacturer's war ranty and may result in structural deficiencies.

Lateral Stability

The design values contained in the evaluation repor assume continuous lateral restraint of the joist's compres sion edge and lateral torsional restraint at the suppo locations. Lateral restraint is generally provided by dia phragm sheathing or bracing spaced at 16" on center less (based on 1½" width joist flanges) nailed to the joist' compression flange.

Applications without continuous lateral bracing wi generally have reduced moment design capacities. Th reduced capacity results from the increased potential fo lateral buckling of the joist's compression flange. Consulta tion with individual manufacturers is recommended for a applications without continuous lateral bracing.

Special Loads or Applications

Wood I-joists are configured and optimized to ac primarily as joists to resist bending loads supported a the bearing by the bottom flange. Applications that resu in significant axial tension or compression loads, requir web holes, special connections, or other unusual cond tions should be evaluated only with the assistance of th individual wood I-joist manufacturers.

Introduction

The wood I-joist is similar to conventional lumber in that it is based on the same raw materials, but differs in how the material is composed. For this reason, conventional lumber design practices are not always compatible with the unique configuration and wood fiber orientation of the wood I-joist. Designers using wood I-joists should develop solutions in accordance with the following guidelines.

Durability issues cannot be overemphasized. See M4.4 for more information about durability.

In addition to detailing for durability, another significant issue in the planning of wood structures is that of fire performance, which is discussed in *NDS* Chapter 16.

Design Span

The design span used for determining critical shears and moments is defined as the clear span between the faces of support plus one-half the minimum required bearing on each end (see Figure M7.4-1). For most wood I-joists, the minimum required end bearing length varies from 1½" to 3½" (adding 2" to the clear span dimension is a good estimate for most applications). At locations of continuity over intermediate bearings, the design span is measured from the centerline of the intermediate support to the face of the bearing at the end support, plus one-half the minimum required bearing length. For interior spans of a continuous joist, the design span extends from centerline to centerline of the intermediate bearings.

Figure M7.4-1   Design Span Determination

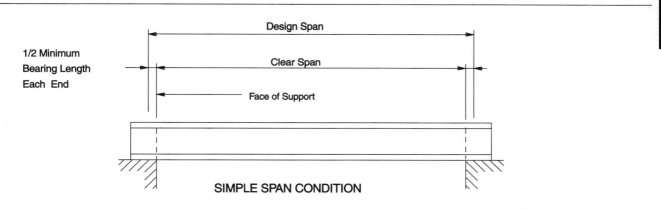

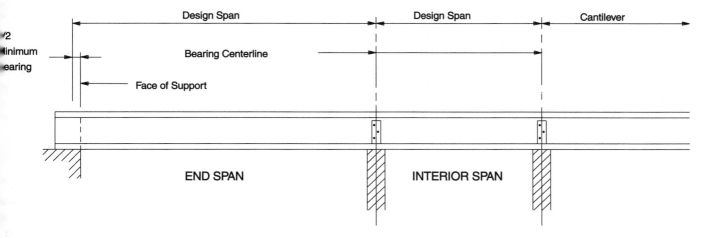

Load Cases

Most building codes require consideration of a critical distribution of loads. Due to the long length and continuous span capabilities of the wood I-joist, these code provisions have particular meaning. Considering a multiple span member, the following design load cases should be considered:

- All spans with total loads
- Alternate span loading
- Adjacent span loading
- Partial span loading (joists with holes)
- Concentrated load provisions (as occurs)

A basic description of each of these load cases follows:

Total loads on all spans – This load case involves placing all live and dead design loads on all spans simultaneously.

Alternate span loading – This load case places the L, L_R, S, or R load portion of the design loads on every other span and can involve two loading patterns. The first pattern results in the removal of the live loads from all even numbered spans. The second pattern removes live loads from all odd numbered spans. For roof applications, some building codes require removal of only a portion of the live loads from odd or even numbered spans. The alternate span load case usually generates maximum end reactions, mid-span moments, and mid-span deflections. Illustrations of this type of loading are shown in Figure M7.4-2.

Adjacent span loading – This load case (see Figure M7.4-2) removes L, L_R, S or R loads from all but two adjoining spans. All other spans, if they exist, are loaded with dead loads only. Depending on the number of spans involved, this load case can lead to a number of load patterns. All combinations of adjacent spans become separate loadings. This load case is used to develop maximum shears and reactions at internal bearing locations.

Partial span loading – This load case involves applying L, L_R, S or R loads to less than the full length of a span (see Figure M7.4-2). For wood I-joists with web holes, this case is used to develop shear at hole locations. When this load case applies, uniform L, L_R, S, R load is applied only from an adjacent bearing to the opposite edge of a rectangular hole (centerline of a circular hole). For each hole within a given span, there are two corresponding load cases. Live loads other than the uniform application load, located within the span containing the hole, are also applied simultaneously. This includes all special loads such as point or tapered loads.

Concentrated load provisions – Most building codes have a concentrated load (live load) provision in addition to standard application design loads. This load case considers this concentrated load to act in combination with the system dead loads on an otherwise unloaded floor or roof. Usually, this provision applies to non-residential construction. An example is the "safe" load applied over a 2½ square foot area for office floors. This load case helps insure the product being evaluated has the required shear and moment capacity throughout it's entire length and should be considered when analyzing the effect of web holes.

A properly designed multiple span member requires numerous load case evaluations. Most wood I-joist manufacturers have developed computer programs, load and span tables, or both that take these various load cases into consideration.

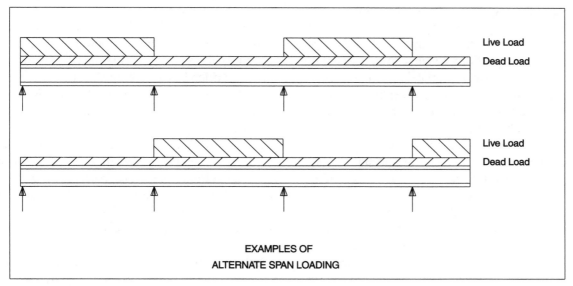

EXAMPLES OF
ALTERNATE SPAN LOADING

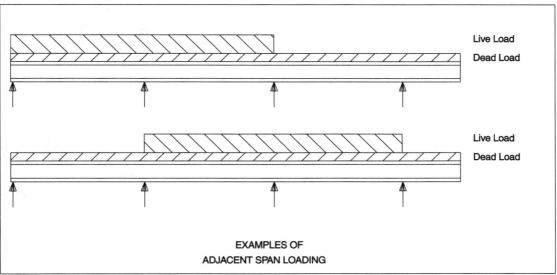

EXAMPLES OF
ADJACENT SPAN LOADING

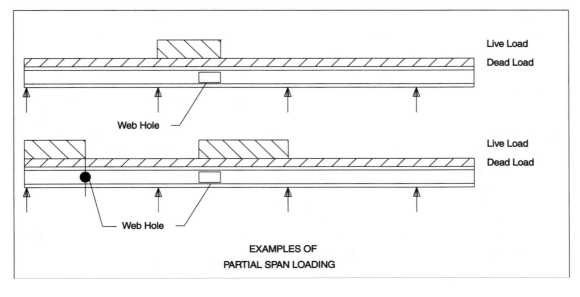

EXAMPLES OF
PARTIAL SPAN LOADING

Floor Performance

Designing a floor system to meet the minimum requirements of a building code may not always provide acceptable performance to the end user. Although minimum criteria help assure a floor system can safely support the imposed loads, the system ultimately must perform to the satisfaction of the end user. Since expectancy levels may vary from one person to another, designing a floor system becomes a subjective issue requiring judgment as to the sensitivity of the intended occupant.

Joist deflection is often used as the primary means for designing floor performance. Although deflection is a factor, there are other equally important variables that can influence the performance of a floor system. A glue-nailed floor system will generally have better deflection performance than a nailed only system. Selection of the decking material is also an important consideration. Deflection of the sheathing material between joists can be reduced by placing the joists at a closer on center spacing or increasing the sheathing thickness.

Proper installation and job site storage are important considerations. All building materials, including wood I-joists, need to be kept dry and protected from exposure to the elements. Proper installation includes correct spacing of sheathing joints, care in fastening of the joists and sheathing, and providing adequate and level supports. All of these considerations are essential for proper system performance.

Vibration may be a design consideration for floor systems that are stiff and where very little dead load (i.e., partition walls, ceilings, furniture, etc.) exists. Vibration can generally be damped with a ceiling system directly attached to the bottom flange of the wood I-joists. Effective bridging or continuous bottom flange nailers (i.e., 2x4 nailed flat-wise and perpendicular to the joist and tied off to the end walls) can also help minimize the potential for vibration in the absence of a direct applied ceiling. Limiting the span/depth ratio of the I-joist may also improve floor performance.

Joist Bearing

Bearing design for wood I-joists requires more than consideration of perpendicular to grain bearing values. Minimum required bearing lengths take into account a number of considerations. These include: cross grain bending and tensile forces in the flanges, web stiffener connection to the joist web, adhesive joint locations and strength, and perpendicular to grain bearing stresses. The model building code evaluation reports provide a source for bearing design information, usually in the form of minimum required bearing lengths.

Usually, published bearing lengths are based on the maximum allowable shear capacity of the particular product and depth or allowable reactions are related to specific bearing lengths. Bearing lengths for wood I-joist are most often based on empirical test results rather than a calculated approach. Each specific manufacturer should be consulted for information when deviations from published criteria are desired.

To better understand the variables involved in a wood I-joist bearing, it's convenient to visualize the member as a composition of pieces, each serving a specific task. For a typical simple span joist, the top flange is a compression member, the bottom flange is a tension member, and the web resists the vertical shear forces. Using this concept, shear forces accumulate in the web member at the bearing locations and must be transferred through the flanges to the support structure. This transfer involves two critical interfaces: between the flange and support member and between the web and flange materials.

Starting with the support member, flange to support bearing involves perpendicular to grain stresses. The lowest design value for either the support member or flange material is usually used to develop the minimum required bearing area.

The second interface to be checked is between the lower joist flange and the bottom edge of the joist web assuming a bottom flange bearing condition. This connection, usually a routed groove in the flange and a matching shaped profile on the web, is a glued joint secured with a waterproof structural adhesive. The contact surface include the sides and bottom of the routed flange.

In most cases, the adhesive line stresses at this joint control the bearing length design. The effective bearing length of the web into the flange is approximately the length of flange bearing onto the support plus an incremental length related to the thickness and stiffness of the flange material.

Since most wood I-joists have web shear capacity in excess of the flange to web joint strength, connection reinforcement is sometimes utilized. The most common method of reinforcement is the addition of web stiffeners (also commonly referred to as bearing blocks). Web stiffeners are vertically oriented wood blocks positioned on both sides of the web. Web stiffeners should be cut so that a gap of at least 1/8" is between the stiffener and the flange to avoid a force fit. Stiffeners are positioned tight to the bottom flange at bearing locations and snug to the bottom of the top flange beneath heavy point loads within a span. Figure M7.4-3 provides an illustration of a typical end bearing assembly.

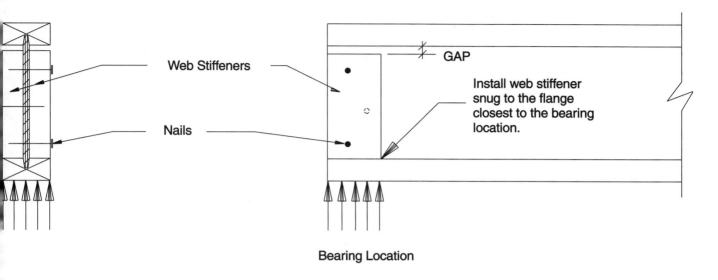

END VIEW   SIDE VIEW

eb Stiffeners

When correctly fastened to the joist web, web stiffeners transfer some of the load from the web into the top of the bottom flange. This reduces the loads on the web to flange joint. A pair of web stiffeners (one on each side) is usually mechanically connected to the web with nails or staples loaded in double shear. For some of the higher capacity wood I-joists, nailing and supplemental gluing with a structural adhesive is required. The added bearing capacity achievable with web stiffeners is limited by the allowable bearing stresses where the stiffeners contact the bearing flange and by their mechanical connection to the web.

Web stiffeners also serve the implied function of reinforcing the web against buckling. Since shear capacity usually increases proportionately with the depth, web stiffeners are very important for deep wood I-joists. For example, a 30" deep wood I-joist may only develop 20% to 30% of its shear and bearing capacity without properly attached web stiffeners at the bearing locations. This is especially important at continuous span bearing locations, where reaction magnitudes can exceed simple span reactions by an additional 25%.

Web stiffeners should be cut so that a gap of at least 1/8" is between the stiffener and the top or bottom of the flange to avoid a force fit. Web stiffeners should be installed snug to the bottom flange for bearing reinforcement or snug to the top flange if under concentrated load from above.

For shallow depth joists, where relatively low shear capacities are required, web stiffeners may not be needed. When larger reaction capacities are required, web stiffener reinforcement may be needed, especially where short bearing lengths are desired. Figure M7.4-4 illustrates the bearing interfaces.

7

M7: PREFABRICATED WOOD I-JOISTS

Figure M7.4-4   Web Stiffener Bearing Interface

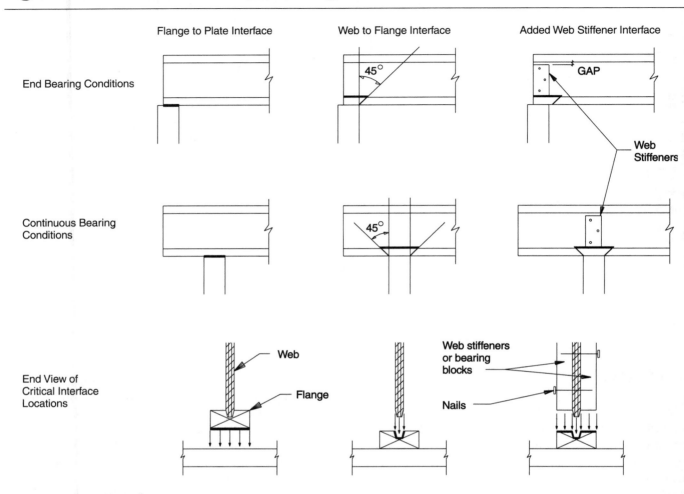

Flange to Plate Interface   Web to Flange Interface   Added Web Stiffener Interface

End Bearing Conditions

GAP

Web Stiffeners

Continuous Bearing Conditions

45°   45°

End View of Critical Interface Locations

Web

Flange

Web stiffeners or bearing blocks

Nails

Beveled End Cuts

Beveled end cuts, where the end of the joist is cut on an angle (top flange does not project over the bearing, much like a fire cut), also requires special design consideration. Again the severity of the angle, web material, location of web section joints, and web stiffener application criteria effect the performance of this type of bearing condition. The specific wood I-joist manufacturers should be consulted for limits on this type of end cut.

It is generally accepted that if a wood I-joist has the minimum required bearing length, and the top flange of the joist is not cut beyond the face of bearing (measured from a line perpendicular to the joist's bottom flange), there is no reduction in shear or reaction capacity. This differs from the conventional lumber provision that suggests there is no decrease in shear strength for beveled cuts of up to an angle of 45°. The reason involves the composite nature of the wood I-joist and how the member fails in shear and or bearing. Figure M7.4-5 provides an illustration of the beveled end cut limitation.

Figure M7.4-5   Beveled End Cut

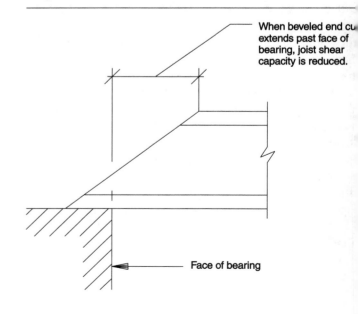

When beveled end cut extends past face of bearing, joist shear capacity is reduced.

Face of bearing

Sloped Bearing Conditions

Sloped bearing conditions require design considerations different from conventional lumber. An example is a birdsmouth bearing cut (notches in the bottom flange, see Figure M7.4-6). This type of bearing should only be used in the low end bearing for wood I-joists. Another example is the use of metal joist support connectors that attach only to the web area of the joist and do not provide a bottom seat in which to bear. In general, this type of connector is not recommended for use with wood I-joists without consideration for the resulting reduced capacity.

The birdsmouth cut is a good solution for the low end bearing when the slope is steep and the tangential loads are high (loads along the axis of the joist member). This assumes the quality of construction is good and the cuts are made correctly and at the right locations. This type of bearing cut requires some skill and is not easy to make, particularly with the wider flange joists. The bearing capacity, especially with high shear capacity members, may be reduced as a result of the cut since the effective flange bearing area is reduced. The notched cut will also reduce the member's shear and moment capacity at a cantilever location.

An alternative to a birdsmouth cut is a beveled bearing plate matching the joist slope or special sloped seat bearing hardware manufactured by some metal connector suppliers. These alternatives also have special design considerations with steep slope applications. As the member slope increases, so does the tangential component of reaction, sometimes requiring additional flange to bearing nailing or straps to provide resistance. Figure M7.4-6 shows some examples of acceptable low end bearing conditions.

Figure M7.4-6   Sloped Bearing Conditions (Low End)

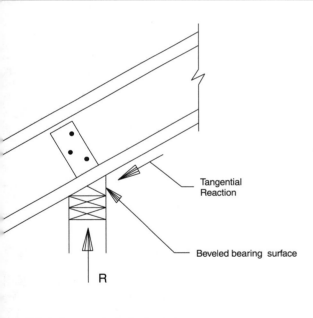

Tangential Reaction

Beveled bearing surface

R

(Tangential reactions may exceed a standard nail connection capacity on steep slopes above 20 to 30 degrees.)

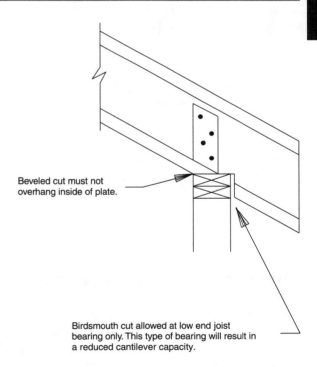

Beveled cut must not overhang inside of plate.

Birdsmouth cut allowed at low end joist bearing only. This type of bearing will result in a reduced cantilever capacity.

For the high end support, bottom flange bearing in a suitable connector or on a beveled plate is recommended. When slopes exceed 30°, straps or gussets may be needed to resist the tangential component of the reaction.

Support connections only to the web area of a wood I-joist, especially at the high end of a sloped application, are not generally recommended. Since a wood I-joist is comprised of a number of pieces, joints between web sections occurring near the end of the member may reduce the joist's shear capacity when not supported from the bottom flange.

When a wood I-joist is supported from the web only the closest web to web joint from the end may be stresse in tension. This could result in a joint failure with th web section pulling out of the bottom flange. Locatin these internal joints away from the end of the member o applying joint reinforcements are potential remedies, bu generally are not practical in the field.

The best bearing solution is to provide direct sup port to the joist's bottom flange to avoid reductions i capacity. Figure M7.4-7 shows typical high end bearin conditions.

Figure M7.4-7   Sloped Bearing Conditions (High End)

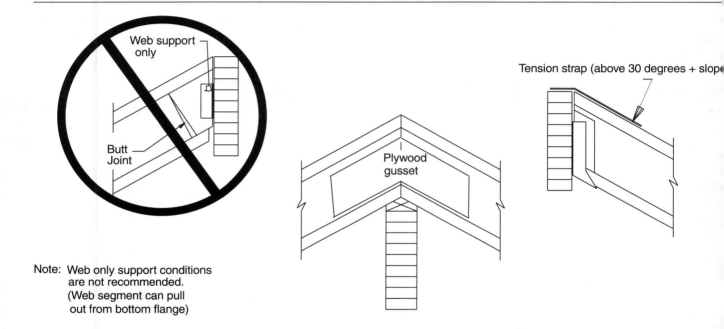

Web support only

Butt Joint

Note: Web only support conditions
are not recommended.
(Web segment can pull
out from bottom flange)

Plywood gusset

Tension strap (above 30 degrees + slope

Beveled bearing surface

Connector Design/Joist Hangers

Although there are numerous hangers and connectors available that are compatible with wood I-joists, many are not. Hangers developed for conventional lumber or glulam beams often use large nails and space them in a pattern that will split the joist flanges and web stiffeners. Hanger selection considerations for wood I-joists should include nail length and diameter, nail location, wood I-joist bearing capacity, composition of the supporting member, physical fit, and load capacity. For example, hangers appropriate for a wood I-joist to glulam beam support may not be compatible for an I-joist to I-joist connection.

In general, nails into the flanges should not exceed the diameter of a 10d common nail, with a recommended length no greater than 1½". Nails into web stiffeners should

not exceed the diameter of a 16d common nail. Nail through the sides of the hanger, when used in combina tion with web stiffeners, can be used to reduce the joist' minimum required bearing length. Nails help transfer load directly from the I-joist web into the hanger, reducin the load transferred through direct bearing in the botton hanger seat.

Hangers should be capable of providing lateral suppo to the top flange of the joist. This is usually accomplishe by a hanger flange that extends the full depth of the jois As a minimum, hanger support should extend to at leas mid-height of a joist used with web stiffeners. Some con nector manufacturers have developed hangers specificall for use with wood I-joists that provide full lateral suppo without the use of web stiffeners. Figure M7.4-8 illustrate lateral joist support requirements for hangers.

Figure M7.4-8 Lateral Support Requirements for Joists in Hangers

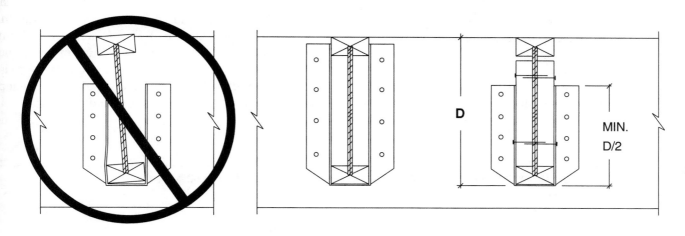

Top and bottom joist flanges must be laterally restrained against rotation.

When top flange style hangers are used to support one I-joist from another, especially the wider flange I-joists, web stiffeners need to be installed tight to the bottom side of the support joist's top flanges. This prevents cross grain bending and rotation of the top flange (see Figure M7.4-9).

When face-nail hangers are used for joist to joist connections, nails into the support joist should extend through and beyond the web element (Figure M7.4-10).

Filler blocks should also be attached sufficiently to provide support for the hanger. Again, nail diameter should be considered to avoid splitting the filler block material.

Multiple I-joists need to be adequately connected together to achieve desired performance. This requires proper selection of a nailing or bolting pattern and attention to web stiffener and blocking needs. Connections should be made through the webs of the I-joists and never through the flanges.

Figure M7.4-9 Top Flange Hanger Support

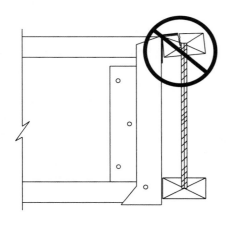

Caution: Large diameter nails can cause splitting.

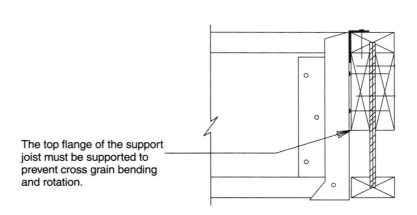

The top flange of the support joist must be supported to prevent cross grain bending and rotation.

For a double I-joist member loaded from one side only, the minimum connection between members should be capable of transferring at least 50% of the applied load. Likewise, for a triple member loaded from one side only, the minimum connection between members must be capable of transferring at least 2/3 of the applied load. The actual connection design should consider the potential slip and differential member stiffness. Many manufacturers recommend limiting multiple members to three joists. Multiple I-joists with 3½" wide flanges may be further limited to two members.

The low torsional resistance of most wood I-joists i also a design consideration for joist to joist connection Eccentrically applied side loads, such as a top flang hanger hung from the side of a double joist, create th potential for joist rotation. Bottom flange restrainin straps, blocking, or directly applied ceiling systems ma be needed on heavily loaded eccentric connections to resis rotation. Figure M7.4-10 shows additional I-joist connec tion considerations for use with face nail hangers.

Figure M7.4-10 Connection Requirements for Face Nail Hangers

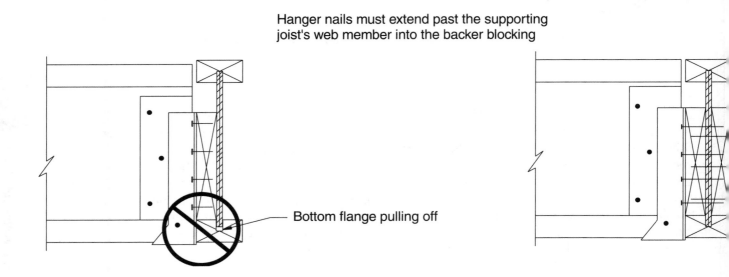

Hanger nails must extend past the supporting joist's web member into the backer blocking

Bottom flange pulling off

Vertical Load Transfer

Bearing loads originating above the joists at the bearing location require blocking to transfer these loads around the wood I-joist to the supporting wall or foundation. This is typically the case in a multi-story structure where bearing walls stack and platform framing is used. Usually, the available bearing capacity of the joist is needed to support its reaction, leaving little if any excess capacity to support additional bearing wall loads from above.

The most common type of blocking uses short pieces of wood I-joist, often referred to as blocking panels, positioned directly over the lower bearing and cut to fit in between the joists. These panels also provide lateral support for the joists and an easy means to transfer lateral diaphragm shears.

The ability to transfer lateral loads (due to wind, seis mic, construction loads, etc.) to shear walls or foundation below is important to the integrity of the building design Compared with dimension lumber blocking, which usuall is toe-nailed to the bearing below, wood I-joist blockin can develop higher diaphragm transfer values because o a wider member width and better nail values.

Specialty products designed specifically for rim board are pre-cut in strips equal to the joist depth and provid support for the loads from above. This solution may als provide diaphragm boundary nailing for lateral loads.

A third method uses vertically oriented short studs, often called squash blocks or cripple blocks, on each side of the joist and cut to a length slightly longer than the depth of the joist. This method should be used in combination with some type of rim joist or blocking material when lateral stability or diaphragm transfer is required.

The use of horizontally oriented sawn lumber as a blocking material is unacceptable. Wood I-joists generally do not shrink in the vertical direction due to their panel type web, creating the potential for a mismatch in height as sawn lumber shrinks to achieve equilibrium. When conventional lumber is used in the vertical orientation, shrinkage problems are not a problem because changes in elongation due to moisture changes are minimal. Figure M7.4-11 shows a few common methods for developing vertical load transfer.

Figure M7.4-11 Details for Vertical Load Transfer

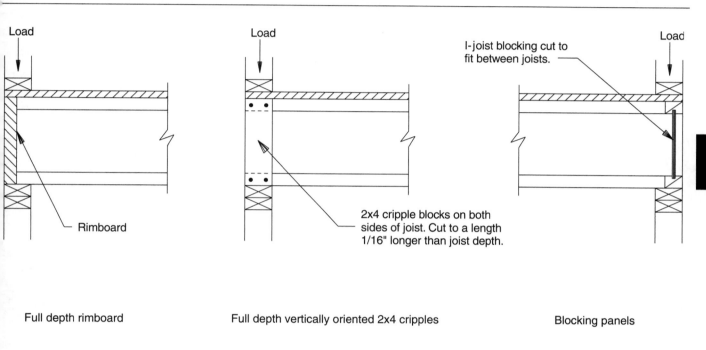

Full depth rimboard

Full depth vertically oriented 2x4 cripples

Blocking panels

Web Holes

Holes cut in the web area of a wood I-joist affect the member's shear capacity. Usually, the larger the hole, the greater the reduction in shear capacity. For this reason, holes are generally located in areas where shear stresses are low. This explains why the largest holes are generally permitted near mid-span of a member. The required spacing between holes and from the end of the member is dependent upon the specific materials and processes used during manufacturing.

The allowable shear capacity of a wood I-joist at a hole location is influenced by a number of variables. These include: percentage of web removed, proximity to a vertical joint between web segments, the strength of the web to flange glue joint, the stiffness of the flange, and the shear strength of the web material. Since wood I-joists are manufactured using different processes and materials, each manufacturer should be consulted for the proper web hole design.

The methodology used to analyze application loads is important in the evaluation of web holes. All load cases that will develop the highest shear at the hole location should be considered. Usually, for members resisting simple uniform design loads, the loading condition that develops the highest shear loads in the center area of a joist span involves partial span loading.

Web holes contribute somewhat to increased deflection. The larger the hole the larger the contribution. Provided not too many holes are involved, the contribution is negligible. In most cases, if the manufacturer's hole criteria are followed and the number of holes is limited to three or less per span, the additional deflection does not warrant consideration.

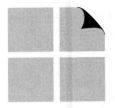

M8: STRUCTURAL COMPOSITE LUMBER

M8.1 General   54

M8.2 Reference Design Values   55

M8.3 Adjustment of Reference
      Design Values   56

M8.4 Special Design Considerations   57

8

M8.1 General

Product Information

Structural composite lumber (SCL) products are well known throughout the construction industry. The advantages of SCL include environmental benefits from better wood fiber utilization along with higher strength, stiffness, and consistency from fiber orientation and manufacturing process control.

SCL is manufactured from strips or full sheets of veneer. The process typically includes alignment of stress graded fiber, application of adhesive, and pressing the material together under heat and pressure. By redistributing natural defects and through state of the art quality control procedures, the resulting material is extremely consistent and maximizes the strength and stiffness of the wood fiber.

The material is typically produced in a long length continuous or fixed press in a billet form. This is then resawn into required dimensions for use. Material is currently available in a variety of depths from 4-3/8" to 24" and thicknesses from 3/4" to 7".

SCL is available in a wide range of sizes and grades. When specifying SCL products, a customer may specify on the basis of size, stress (strength), or appearance.

SCL products are proprietary and are covered by code acceptance reports by one or all of the model building codes. Such reports should be consulted for current design information while manufacturer's literature can be consulted for design information, sizing tables, and installation recommendations.

Common Uses

SCL is widely used as a framing material for housing. SCL is made in different grades and with various processes and can be utilized in numerous applications. Proper design is required to optimize performance and economics.

In addition to use in housing, SCL finds increasing use in commercial and industrial construction. Its high strength, stiffness, universal availability, and cost saving attributes make it a viable alternative in most low-rise construction projects.

SCL is used as beams, headers, joists, rafters, studs and plates in conventional construction. In addition, SCL used to fabricate structural glued laminated beams, trusses and prefabricated wood I-joists.

Availability

SCL is regarded as a premium construction material and is widely available. To efficiently specify SCL for individual projects, the customer should be aware of the species and strength availability. Sizes vary with each individual product. The best source of this information is your local lumber supplier, distribution center, or SCL manufacturer. Proper design is facilitated through the use of manufacturer's literature, code reports, and software available from SCL manufacturers.

M8.2 Reference Design Values

General

As stated in *NDS* 8.2, SCL products are proprietary and each manufacturer develops design values appropriate for their products. These values are reviewed by the model building codes and published in acceptance reports and manufacturer's literature.

Reference design values are used in conjunction with the adjustment factors in M8.3.

Shear Design

SCL is typically designed and installed as a rectangular section. Loads near supports may be reduced per *NDS* 3.4.3.1. However, such load must be included in bearing calculations. Shear values for SCL products often change with member orientation.

Bearing

SCL typically has high F_c and $F_{c\perp}$ properties. With the higher shear and bending capacities, shorter or continuous spans are often controlled by bearing. The user is cautioned to ensure the design accounts for compression of the support material (i.e., plate) as well as the beam material. Often the plate material is of softer species and will control the design.

Bending

Published bending capacities of SCL beams are determined from testing of production specimens. Adjustment for the size of the member is also determined by test.

Field notching or drilling of holes is typically not allowed. Similarly, excessive nailing or the use of improper nail sizes can cause splitting that will also reduce capacity. The manufacturer should be contacted when evaluating a damaged beam.

Deflection Design

Deflection calculations for SCL typically are similar to provisions for other rectangular wood products (see M3.5). Values for use in deflection equations can be found in the individual manufacturer's product literature or evaluation reports. Some manufacturers might publish "true" E values which would require additional calculations to account for shear deflection (see *NDS* Appendix F).

8

M8: STRUCTURAL COMPOSITE LUMBER

M8.3 Adjustment of Reference Design Values

Member design capacity is the product of reference design values, adjustment factors, and section properties. Reference design values for SCL are discussed in M8.2.

Adjustment factors are provided for applications outside the reference end-use conditions and for member configuration effects as specified in *NDS* 8.3. When one or more of the specific end use or member configuration conditions are beyond the range of the reference conditions, these adjustment factors shall be used to modify the appropriate property. Adjustment factors for the effects of moisture, temperature, member configuration, and size are provided in *NDS* 8.3. Additional adjustment factors can be found in the manufacturer's product literature or code evaluation report. Table M8.3-1 shows the applicability of adjustment factors for SCL in a slightly different format for the designer.

Certain products may not be suitable for use in some applications or with certain treatments. Such conditions can result in structural deficiencies and may void manufacturer warranties. The manufacturer or code evaluation report should be consulted for specific information.

Table M8.3-1 Applicability of Adjustment Factors for Structural Composite Lumber[1]

Allowable Stress Design	Load and Resistance Factor Design
$F_b' = F_b\ C_D\ C_M\ C_t\ C_L\ C_V\ C_r$	$F_b' = F_b\ C_M\ C_t\ C_L\ C_V\ C_r\ K_F\ \phi_b\ \lambda$
$F_t' = F_t\ C_D\ C_M\ C_t$	$F_t' = F_t\ C_M\ C_t\ K_F\ \phi_t\ \lambda$
$F_v' = F_v\ C_D\ C_M\ C_t$	$F_v' = F_v\ C_M\ C_t\ K_F\ \phi_v\ \lambda$
$F_{c\perp}' = F_{c\perp}\ C_M\ C_t\ C_b$	$F_{c\perp}' = F_{c\perp}\ C_M\ C_t\ C_b\ K_F\ \phi_c\ \lambda$
$F_c' = F_c\ C_D\ C_M\ C_t\ C_P$	$F_c' = F_c\ C_M\ C_t\ C_P\ K_F\ \phi_c\ \lambda$
$E' = E\ C_M\ C_t$	$E' = E\ C_M\ C_t$
$E_{min}' = E_{min}\ C_M\ C_t$	$E_{min}' = E_{min}\ C_M\ C_t\ K_F\ \phi_s$

1. See *NDS* 8.3.6 for information on simultaneous application of the volume factor, C_V, and the beam stability factor, C_L.

Bending Member Example

For fully laterally supported members stressed in strong axis bending and used in a normal building environment (meeting the reference conditions of *NDS* 2.3 and 8.3), the adjusted design values reduce to:

For ASD:

$$F_b' = F_b\ C_D\ C_V$$
$$F_v' = F_v\ C_D$$
$$F_{c\perp}' = F_{c\perp}\ C_b$$
$$E' = E$$

For LRFD:

$$F_b' = F_b\ C_V\ K_F\ \phi_b\ \lambda$$
$$F_v' = F_v\ K_F\ \phi_v\ \lambda$$
$$F_{c\perp}' = F_{c\perp}\ C_b\ K_F\ \phi_c\ \lambda$$
$$E' = E$$

Axially Loaded Member Example

For axially loaded members used in a normal building environment (meeting the reference conditions of *NDS* 2.3 and 8.3) designed to resist tension or compression loads, the adjusted tension or compression design values reduce to:

For ASD:

$$F_c' = F_c\ C_D\ C_P$$
$$F_t' = F_t\ C_D$$
$$E_{min}' = E_{min}$$

For LRFD:

$$F_c' = F_c\ C_P\ K_F\ \phi_c\ \lambda$$
$$F_t' = F_t\ K_F\ \phi_t\ \lambda$$
$$E_{min}' = E_{min}\ K_F\ \phi_s$$

M8.4 Special Design Considerations

General

With proper detailing and protection, SCL can perform well in a variety of environments. One key to proper detailing is planning for the natural shrinkage and swelling of wood members as they are subjected to various drying and wetting cycles. While moisture changes have the largest impact on lumber dimensions, some designs must also check the effects of temperature. While SCL is typically produced using dry veneer, some moisture accumulation may occur during storage. If the product varies significantly from specified dimensions, the user is cautioned from using such product as it will "shrink" as it dries.

In addition to designing to accommodate dimensional changes and detailing for durability, another significant issue in the planning of wood structures is that of fire performance, which is covered in Chapter M16.

Dimensional Changes

The dimensional stability and response to temperature effects of engineered lumber is similar to that of solid sawn lumber of the same species.

Some densification of the wood fiber can occur in various manufacturing processes. SCL that is densified will result in a product that has more wood fiber in a given volume and can therefore hold more water than a solid sawn equivalent. When soaked these products expand and dimensional changes can occur.

Adhesive applied during certain processes tends to form a barrier to moisture penetration. Therefore, the material will typically take longer to reach equilibrium than its solid sawn counterpart.

For given temperatures and applications, different levels of relative humidity are present. This will cause the material to move toward an equilibrium moisture content (EMC). Eventually all wood products will reach their EMC for a given environment. SCL will typically equilibrate at a lower EMC (typically 3% to 4% lower) than solid sawn lumber and will take longer to reach an ambient EMC. Normal swings in humidity during the service life of the structure should not produce noticeable dimensional changes in SCL members.

More information on designing for moisture and temperature change is included in M4.4.

Durability

Designing for durability is a key part of the architectural and engineering design of the building. Wood exposed to high levels of moisture can decay over time. While there are exceptions – such as naturally durable species, preservative-treated wood, and those locations that can completely air-dry between moisture cycles – prudent design calls for a continuing awareness of the possibility of moisture accumulation. Awareness of the potential for decay is the key – many design conditions can be detailed to minimize the accumulation of moisture; for other problem conditions, preservative-treated wood or naturally durable species should be specified.

This section cannot cover the topic of designing for durability in detail. There are many excellent texts that devote entire chapters to the topic, and designers are advised to use this information to assist in designing "difficult" design areas, such as:

- structures in high moisture or humid conditions
- where wood comes in contact with concrete or masonry
- where wood members are supported in steel hangers or connectors in which condensation could collect
- anywhere that wood is directly or indirectly exposed to the elements
- where wood, if it should ever become wet, could not naturally dry out

This list is not intended to be all-inclusive – it is merely an attempt to alert designers to special conditions that may cause problems when durability is not considered in the design.

More information on detailing for durability is included in M4.4.

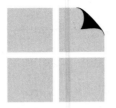

M9: WOOD STRUCTURAL PANELS

M9.1	General	60
M9.2	Reference Design Values	60
M9.3	Adjustment of Reference Design Values	66
M9.4	Special Design Considerations	67

9

M9.1 General

Product Description

Wood structural panels are wood-based panel products that have been rated for use in structural applications. Common applications for wood structural panels include roof sheathing, wall sheathing, subflooring, and single-layer flooring (combination subfloor-underlayment). Plywood is also manufactured in various sanded grades.

Wood structural panels are classified by span ratings. Panel span ratings identify the maximum recommended support spacings for specific end uses. Design capacities are provided on the basis of span ratings.

Sanded grades are classed according to nominal thickness and design capacities are provided on that basis.

Designers must specify wood structural panels by the span ratings, nominal thicknesses, grades, and constructions associated with tabulated design recommendations. Exposure durability classification must also be identified.

Single Floor panels may have tongue-and-groove or square edges. If square edge Single Floor panels are specified, the specification shall require lumber blocking between supports.

Table M9.1-1 provides descriptions and typical uses for various panel grades and types.

M9.2 Reference Design Values

General

Wood structural panel design capacities listed in Tables M9.2-1 through M9.2-2 are minimum for grade and span rating. Multipliers shown in each table provide adjustments in capacity for Structural I panel grades. To take advantage of these multipliers, the specifier must insure that the correct panel is used in construction.

The tabulated capacities and adjustment factors are based on data from tests of panels manufactured in accordance with industry standards and which bear the trademark of a qualified inspection and testing agency.

Structural panels have a strength axis direction and a cross panel direction. The direction of the strength axis is defined as the axis parallel to the orientation of OSB face strands or plywood face veneer grain and is the long dimension of the panel unless otherwise indicated by the manufacturer. This is illustrated in Figure M9.2-1.

Panel Stiffness and Strength

Panel design capacities listed in Table M9.2-1 are based on flat panel bending (Figure M9.2-2) as measured by testing according to principles of ASTM D3043 Method C (large panel testing).

Stiffness (EI)
Panel bending stiffness is the capacity to resist deflection and is represented as EI. E is the reference modulus of elasticity of the material, and I is the moment of inertia of the cross section. The units of EI are lb-in.^{2} per foot of panel width.

Strength (F_bS)
Bending strength capacity is the design maximum moment, represented as F_bS. F_b is the reference extreme fiber bending stress of the material, and S is the section modulus of the cross section. The units of F_bS are lb-in. per foot of panel width.

Figure M9.2-1 Structural Panel with Strength Direction Across Supports

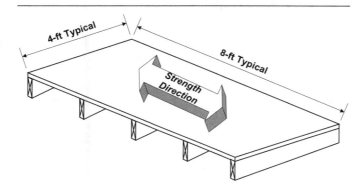

Figure M9.2-2 Example of Structural Panel in Bending

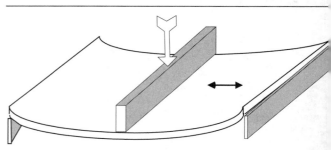

Table M9.1-1   Guide to Panel Use

Panel Grade	Description & Use	Common Nominal Thickness (in.)	Panel Construction		
			OSB	COM-PLY	Plywood & Veneer Grade
Sheathing EXP 1	Unsanded sheathing grade for wall, roof, subflooring, and industrial applications such as pallets and for engineering design with proper capacities. Manufactured with intermediate and exterior glue. For long-term exposure to weather or moisture, only Exterior type plywood is suitable.	5/16, 3/8, 15/32, 1/2, 19/32, 5/8, 23/32, 3/4	Yes	Yes	Yes, face C, back D, inner D
Structural I Sheathing EXP 1	Panel grades to use where shear and cross-panel strength properties are of maximum importance. Made with exterior glue only. Plywood Structural I is made from all Group 1 woods.	19/32, 5/8, 23/32, 3/4	Yes	Yes	Yes, face C, back D, inner D
Single Floor EXP 1	Combination subfloor-underlayment. Provides smooth surface for application of carpet and pad. Possesses high concentrated and impact load resistance during construction and occupancy. Manufactured with intermediate (plywood) and exterior glue. Touch-sanded. Available with tongue-and-groove edges.	19/32, 5/8, 23/32, 3/4, 7/8, 1, 1-3/32, 1-1/8	Yes	Yes	Yes, face C-Plugged, back D, inner D
Underlayment EXP 1 or INT	For underlayment under carpet and pad. Available with exterior glue. Touch-sanded. Available with tongue-and-groove edges.	1/4, 11/32, 3/8, 15/32, 1/2, 19/32, 5/8, 23/32, 3/4	No	No	Yes, face C-Plugged, back D, inner D
C-D-Plugged EXP 1	For built-ins, wall and ceiling tile backing. Not for underlayment. Available with exterior glue. Touch-sanded.	1/2, 19/32, 5/8, 23/32, 3/4	No	No	Yes, face C-Plugged, back D, inner D
Sanded Grades EXP 1 or INT	Generally applied where a high-quality surface is required. Includes APA N-N, N-A, N-B, N-D, A-A, A-D, B-B, and B-D INT grades.	1/4, 11/32, 3/8, 15/32, 1/2, 19/32, 5/8, 23/32, 3/4	No	No	Yes, face B or better, back D or better, inner C & D
Marine EXT	Superior Exterior-type plywood made only with Douglas-fir or western larch. Special solid-core construction. Available with medium density overlay (MDO) or high density overlay (HDO) face. Ideal for boat hull construction.	1/4, 11/32, 3/8, 15/32, 1/2, 19/32, 5/8, 23/32, 3/4	No	No	Yes, face A or face B, back A or inner B

9

M9: WOOD STRUCTURAL PANELS

Table M9.2-1 Wood Structural Panel Bending Stiffness and Strength

Span Rating	Stress Parallel to Strength Axis[1]				Stress Perpendicular to Strength Axis[1]			
	Plywood			OSB	Plywood			OSB
	3-ply	4-ply	5-ply		3-ply	4-ply	5-ply	
PANEL BENDING STIFFNESS, EI (lb-in.^{2}/ft of panel width)								
24/0	66,000	66,000	66,000	60,000	3,600	7,900	11,000	11,000
24/16	86,000	86,000	86,000	78,000	5,200	11,500	16,000	16,000
32/16	125,000	125,000	125,000	115,000	8,100	18,000	25,000	25,000
40/20	250,000	250,000	250,000	225,000	18,000	39,500	56,000	56,000
48/24	440,000	440,000	440,000	400,000	29,500	65,000	91,500	91,500
16oc	165,000	165,000	165,000	150,000	11,000	24,000	34,000	34,000
20oc	230,000	230,000	230,000	210,000	13,000	28,500	40,500	40,500
24oc	330,000	330,000	330,000	300,000	26,000	57,000	80,500	80,500
32oc	715,000	715,000	715,000	650,000	75,000	165,000	235,000	235,000
48oc	1,265,000	1,265,000	1,265,000	1,150,000	160,000	350,000	495,000	495,000
Multiplier for Structural I Panels	1.0	1.0	1.0	1.0	1.5	1.5	1.6	1.6
PANEL BENDING STRENGTH, F$_b$S (lb-in./ft of panel width)								
24/0	250	275	300	300	54	65	97	97
24/16	320	350	385	385	64	77	115	115
32/16	370	405	445	445	92	110	165	165
40/20	625	690	750	750	150	180	270	270
48/24	845	930	1,000	1,000	225	270	405	405
16oc	415	455	500	500	100	120	180	180
20oc	480	530	575	575	140	170	250	250
24oc	640	705	770	770	215	260	385	385
32oc	870	955	1,050	1,050	380	455	685	685
48oc	1,600	1,750	1,900	1,900	680	815	1,200	1,200
Multiplier for Structural I Panels	1.0	1.0	1.0	1.0	1.3	1.4	1.5	1.5

1. Strength axis is defined as the axis parallel to the face and back orientation of the flakes or the grain (veneer), which is generally the long panel direction, unless otherwise marked.

Axial Capacities

Axial Stiffness (EA)

Panel axial stiffnesses listed in Table M9.2-2 are based on testing according to the principles of ASTM D3501 Method B. Axial stiffness is the capacity to resist axial strain and is represented as EA. E is the reference axial modulus of elasticity of the material, and A is the area of the cross section. The units of EA are pounds per foot of panel width.

Tension (F$_t$A)

Tension capacities listed in Table M9.2-2 are based on testing according to the principles of ASTM D350 Method B. Tension capacity is given as F$_t$A. F$_t$ is the reference tensile stress of the material, and A is the area of the cross section. The units of F$_t$A are pounds per foot of panel width.

Table M9.2-2   Wood Structural Panel Axial Stiffness, Tension, and Compression Capacities

| Span Rating | Stress Parallel to Strength Axis[1] | | | | Stress Perpendicular to Strength Axis[1] | | | |
| | Plywood | | | OSB | Plywood | | | OSB |
	3-ply	4-ply	5-ply		3-ply	4-ply	5-ply	
PANEL TENSION, F_tA (lb/ft of panel width)								
24/0	2,300	2,300	3,000	2,300	600	600	780	780
24/16	2,600	2,600	3,400	2,600	990	990	1,300	1,300
32/16	2,800	2,800	3,650	2,800	1,250	1,250	1,650	1,650
40/20	2,900	2,900	3,750	2,900	1,600	1,600	2,100	2,100
48/24	4,000	4,000	5,200	4,000	1,950	1,950	2,550	2,550
16oc	2,600	2,600	3,400	2,600	1,450	1,450	1,900	1,900
20oc	2,900	2,900	3,750	2,900	1,600	1,600	2,100	2,100
24oc	3,350	3,350	4,350	3,350	1,950	1,950	2,550	2,550
32oc	4,000	4,000	5,200	4,000	2,500	2,500	3,250	3,250
48oc	5,600	5,600	7,300	5,600	3,650	3,650	4,750	4,750
Multiplier for Structural I Panels	1.0	1.0	1.0	1.0	1.0	1.0	1.0	1.0
PANEL COMPRESSION, F_cA (lb/ft of panel width)								
24/0	2,850	4,300	4,300	2,850	2,500	3,750	3,750	2,500
24/16	3,250	4,900	4,900	3,250	2,500	3,750	3,750	2,500
32/16	3,550	5,350	5,350	3,550	3,100	4,650	4,650	3,100
40/20	4,200	6,300	6,300	4,200	4,000	6,000	6,000	4,000
48/24	5,000	7,500	7,500	5,000	4,800	7,200	7,200	4,300
16oc	4,000	6,000	6,000	4,000	3,600	5,400	5,400	3,600
20oc	4,200	6,300	6,300	4,200	4,000	6,000	6,000	4,000
24oc	5,000	7,500	7,500	5,000	4,800	7,200	7,200	4,300
32oc	6,300	9,450	9,450	6,300	6,200	9,300	9,300	6,200
48oc	8,100	12,150	12,150	8,100	6,750	10,800	10,800	6,750
Multiplier for Structural I Panels	1.0	1.0	1.0	1.0	1.0	1.0	1.0	1.0
PANEL AXIAL STIFFNESS, EA (lb/ft of panel width)								
24/0	3,350,000	3,350,000	3,350,000	3,350,000	2,900,000	2,900,000	2,900,000	2,900,000
24/16	3,800,000	3,800,000	3,800,000	3,800,000	2,900,000	2,900,000	2,900,000	2,900,000
32/16	4,150,000	4,150,000	4,150,000	4,150,000	3,600,000	3,600,000	3,600,000	3,600,000
40/20	5,000,000	5,000,000	5,000,000	5,000,000	4,500,000	4,500,000	4,500,000	4,500,000
48/24	5,850,000	5,850,000	5,850,000	5,850,000	5,000,000	5,000,000	5,000,000	4,500,000
16oc	4,500,000	4,500,000	4,500,000	4,500,000	4,200,000	4,200,000	4,200,000	4,200,000
20oc	5,000,000	5,000,000	5,000,000	5,000,000	4,500,000	4,500,000	4,500,000	4,500,000
24oc	5,850,000	5,850,000	5,850,000	5,850,000	5,000,000	5,000,000	5,000,000	4,500,000
32oc	7,500,000	7,500,000	7,500,000	7,500,000	7,300,000	7,300,000	7,300,000	5,850,000
48oc	8,200,000	8,200,000	8,200,000	8,200,000	7,300,000	7,300,000	7,300,000	7,300,000
Multiplier for Structural I Panels	1.0	1.0	1.0	1.0	1.0	1.0	1.0	1.0

1. Strength axis is defined as the axis parallel to the face and back orientation of the flakes or the grain (veneer), which is generally the long panel direction, unless otherwise marked.

Compression (F_cA)

Compression (Figure M9.2-3) capacities listed in Table M9.2-2 are based on testing according to the principles of ASTM D3501 Method B. Compressive properties are generally influenced by buckling; however, this effect was eliminated by restraining the edges of the specimens during testing. Compression capacity is given as F_cA. F_c is the reference compression stress of the material, and A is the area of the cross section. The units of F_cA are pounds per foot of panel width.

Shear Capacities

Planar (Rolling) Shear ($F_s[Ib/Q]$)

Shear-in-the-plane of the panel (rolling shear) capacities listed in Table M9.2-3 are based on testing according to the principles of ASTM D2718. Shear strength in the plane of the panel is the capacity to resist horizontal shear breaking loads when loads are applied or developed on opposite faces of the panel (Figure M9.2-4), as in flat panel bending. Planar shear capacity is given as $F_s[Ib/Q]$. F_s is the reference material stress, and Ib/Q is the panel cross-sectional shear constant. The units of $F_s[Ib/Q]$ are pounds per foot of panel width.

Rigidity Through-the-Thickness (G_vt_v)

Panel rigidities listed in Table M9.2-4 are based on testing according to the principles of ASTM D2719 Method C. Panel rigidity is the capacity to resist deformation under shear through the thickness stress (Figure M9.2-5). Rigidity is given as G_vt_v. G_v is the reference modulus of rigidity, and t_v is the effective panel thickness for shear. The units of G_vt_v are pounds per inch of panel depth (for vertical applications). Multiplication of G_vt_v by panel depth gives GA, used by designers for some applications.

Through-the-Thickness Shear (F_vt_v)

Through-the-thickness shear capacities listed in Table M9.2-4 are based on testing according to the principles of ASTM D2719 Method C. Allowable through the thickness shear is the capacity to resist horizontal shear breaking loads when loads are applied or developed on opposite edges of the panel (Figure M9.2-5), such as in an I-beam. Where additional support is not provided to prevent bucking, design capacities in Table M9.2-4 are limited to sections 2 ft or less in depth. Deeper sections may require additional reductions. F_v is the reference stress of the material, and t_v is the effective panel thickness for shear. The units of F_vt_v are pounds per inch of shear resisting panel length.

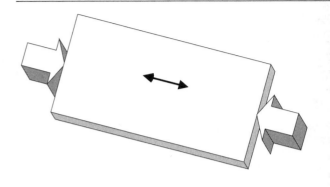

Figure M9.2-3   Structural Panel with Axial Compression Load in the Plane of the Panel

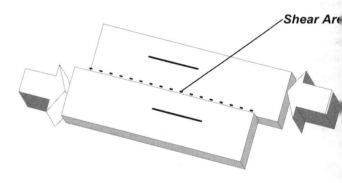

Figure M9.2-4   Through-the-Thickness Shear for Wood Structural Panels

Through-the-Thickness Shear

Shear Area

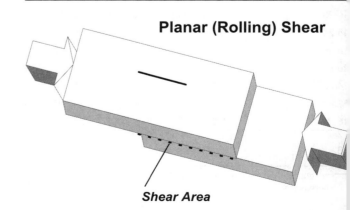

Figure M9.2-5   Planar (Rolling) Shear or Shear-in-the-Plane for Wood Structural Panels

Planar (Rolling) Shear

Shear Area

Table M9.2-3   Wood Structural Panel Planar (Rolling) Shear Capacities

| Span Rating | Stress Parallel to Strength Axis | | | | Stress Perpendicular to Strength Axis | | | |
| | Plywood | | | OSB | Plywood | | | OSB |
	3-ply	4-ply	5-ply		3-ply	4-ply	5-ply	
	PANEL SHEAR-IN-THE-PLANE, F_s(lb/Q) (lb/ft of panel width)							
24/0	155	155	170	130	275	375	130	130
24/16	180	180	195	150	315	435	150	150
32/16	200	200	215	165	345	480	165	165
40/20	245	245	265	205	430	595	205	205
48/24	300	300	325	250	525	725	250	250
16oc	245	245	265	205	430	595	205	205
20oc	245	245	265	205	430	595	205	205
24oc	300	300	325	250	525	725	250	250
32oc	360	360	390	300	630	870	300	300
48oc	460	460	500	385	810	1,100	385	385
Multiplier for Structural I Panels	1.4	1.4	1.4	1.0	1.4	1.4	1.0	1.0

Table M9.2-4   Wood Structural Panel Rigidity and Through-the-Thickness Shear Capacities

| Span Rating | Stress Parallel to Strength Axis | | | | Stress Perpendicular to Strength Axis | | | |
| | Plywood | | | OSB | Plywood | | | OSB |
	3-ply	4-ply	5-ply[1]		3-ply	4-ply	5-ply[1]	
	PANEL RIGIDITY THROUGH-THE-THICKNESS, $G_v t_v$ (lb/in. of panel depth)							
24/0	25,000	32,500	37,500	77,500	25,000	32,500	37,500	77,500
24/16	27,000	35,000	40,500	83,500	27,000	35,000	40,500	83,500
32/16	27,000	35,000	40,500	83,500	27,000	35,000	40,500	83,500
40/20	28,500	37,000	43,000	88,500	28,500	37,000	43,000	88,500
48/24	31,000	40,500	46,500	96,000	31,000	40,500	46,500	96,000
16oc	27,000	35,000	40,500	83,500	27,000	35,000	40,500	83,500
20oc	28,000	36,500	42,000	87,000	28,000	36,500	42,000	87,000
24oc	30,000	39,000	45,000	93,000	30,000	39,000	45,000	93,000
32oc	36,000	47,000	54,000	110,000	36,000	47,000	54,000	110,000
48oc	50,500	65,500	76,000	155,000	50,500	65,500	76,000	155,000
Multiplier for Structural I Panels	1.3	1.3	1.1	1.0	1.3	1.3	1.1	1.0
	PANEL THROUGH-THE-THICKNESS SHEAR, $F_v t_v$ (lb/in. of shear-resisting panel length)							
24/0	53	69	80	155	53	69	80	155
24/16	57	74	86	165	57	74	86	165
32/16	62	81	93	180	62	81	93	180
40/20	68	88	100	195	68	88	100	195
48/24	75	98	115	220	75	98	115	220
16oc	58	75	87	170	58	75	87	170
20oc	67	87	100	195	67	87	100	195
24oc	74	96	110	215	74	96	110	215
32oc	80	105	120	230	80	105	120	230
48oc	105	135	160	305	105	135	160	305
Multiplier for Structural I Panels	1.3	1.3	1.1	1.0	1.3	1.3	1.1	1.0

1. 5-ply applies to plywood with five or more layers. For 5-ply plywood with three layers, use $G_v t_v$ values for 4-ply panels.

9

M9: WOOD STRUCTURAL PANELS

M9.3 Adjustment of Reference Design Values

General

Adjusted panel design capacities are determined by multiplying reference capacities, as given in Tables M9.2-1 through M9.2-4, by the adjustment factors in *NDS* 9.3. Some adjustment factors should be obtained from the manufacturer or other approved source. In the *NDS Commentary*, C9.3 provides additional information on typical adjustment factors.

Tabulated capacities provided in this Chapter are suitable for reference end-use conditions. Reference end-use conditions are consistent with conditions typically associated with light-frame construction. For wood structural panels, these typical conditions involve the use of full-sized untreated panels in moderate temperature and moisture exposures.

Appropriate adjustment factors are provided for applications in which the conditions of use are inconsistent with reference conditions. In addition to temperature and moisture, this includes consideration of panel treatment and size effects.

NDS Table 9.3.1 lists applicability of adjustment factors for wood structural panels. Table M9.3-1 shows the applicability of adjustment factors for wood structural panels in a slightly different format for the designer.

Table M9.3-1   Applicability of Adjustment Factors for Wood Structural Panels

Allowable Stress Design	Load and Resistance Factor Design
$F_bS' = F_bS\ C_D\ C_M\ C_t\ C_G\ C_s$	$F_bS' = F_bS\ C_M\ C_t\ C_G\ C_s\ K_F\ \phi_b\ \lambda$
$F_tA' = F_tA\ C_D\ C_M\ C_t\ C_G\ C_s$	$F_tA' = F_tA\ C_M\ C_t\ C_G\ C_s\ K_F\ \phi_t\ \lambda$
$F_vt_v' = F_vt_v\ C_D\ C_M\ C_t\ C_G$	$F_vt_v' = F_vt_v\ C_M\ C_t\ C_G\ K_F\ \phi_v\ \lambda$
$F_s(Ib/Q)' = F_s(Ib/Q)\ C_D\ C_M\ C_t\ C_G$	$F_s(Ib/Q)' = F_s(Ib/Q)\ C_M\ C_t\ C_G\ K_F\ \phi_v\ \lambda$
$F_cA' = F_cA\ C_D\ C_M\ C_t\ C_G$	$F_cA' = F_cA\ C_M\ C_t\ C_G\ K_F\ \phi_c\ \lambda$
$EI' = EI\ C_M\ C_t\ C_G$	$EI' = EI\ C_M\ C_t\ C_G$
$EA' = EA\ C_M\ C_t\ C_G$	$EA' = EA\ C_M\ C_t\ C_G$
$G_vt_v' = G_vt_v\ C_M\ C_t\ C_G$	$G_vt_v' = G_vt_v\ C_M\ C_t\ C_G$
$F_{c\perp}' = F_{c\perp}\ C_M\ C_t\ C_G$	$F_{c\perp}' = F_{c\perp}\ C_M\ C_t\ C_G\ K_F\ \phi_c\ \lambda$

Bending Member Example

For non-Structural I grade wood structural panels, greater than 24" in width, loaded in bending, and used in a normal building environment (meeting the reference conditions of *NDS* 2.3 and 9.3), the adjusted design values reduce to:

For ASD:

$$F_bS' = F_bS\ C_D$$

$$EI' = EI$$

For LRFD:

$$F_bS' = F_bS\ K_F\ \phi_b\ \lambda$$

$$EI' = EI$$

Axially Loaded Member Example

For non-Structural I grade wood structural panels, greater than 24" in width, axially loaded, and used in a normal building environment (meeting the reference conditions of *NDS* 2.3 and 4.3) designed to resist tension or compression loads, the adjusted tension or compression design values reduce to:

For ASD:

$$F_cA' = F_cA\ C_D$$

$$F_tA' = F_tA\ C_D$$

$$EA' = EA$$

For LRFD:

$$F_cA' = F_cA\ K_F\ \phi_c\ \lambda$$

$$F_tA' = F_tA\ K_F\ \phi_t\ \lambda$$

$$EA' = EA$$

Preservative Treatment

Capacities given in Tables M9.2-1 through M9.2-4 apply without adjustment to plywood pressure-impregnated with preservatives and redried in accordance with American Wood-Preservers' Association (AWPA) Specification C-9 or Specification C-22. However, due to the absence of applicable treating industry standards, OSB and COMPLY panels are not currently recommended for applications requiring pressure-preservative treating.

Fire Retardant Treatment

The information provided in this Chapter does not apply to fire-retardant-treated panels. All capacities and end-use conditions for fire-retardant-treated panels shall be in accordance with the recommendations of the company providing the treating and redrying service.

M9.4 Special Design Considerations

Panel Edge Support

For certain span ratings, the maximum recommended roof span for sheathing panels is dependent upon panel edge support. Although edge support may be provided by lumber blocking, panel clips are typically used when edge support is required. Table M9.4-1 summarizes the relationship between panel edge support and maximum recommended spans.

Table M9.4-1   Panel Edge Support

Sheathing Span Rating	Maximum Recommended Span (in.)	
	With Edge Support	Without Edge Support
24/0	24	20[1]
24/16	24	24
32/16	32	28
40/20	40	32
48/24	48	36

1. 20 in. for 3/8-in. and 7/16-in. panels, 24 in. for 15/32-in. and 1/2-in. panels.

Long-Term Loading

Wood-based panels under constant load will creep (deflection will increase) over time. For typical construction applications, panels are not normally under constant load and, accordingly, creep need not be considered in design. When panels will sustain permanent loads which will stress the product to one-half or more of its design capacity, allowance should be made for creep. Appropriate adjustments should be obtained from the manufacturer or an approved source.

Panel Spacing

Wood-based panel products expand and contract slightly as a natural response to changes in panel moisture content. To provide for in-plane dimensional changes, panels should be installed with a 1/8" spacing at all panel end and edge joints. A standard 10d box nail may be used to check panel edge and panel end spacing.

Minimum Nailing

Minimum nailing for wood structural panel applications is shown in Table M9.4-2.

9

M9: WOOD STRUCTURAL PANELS

Table M9.4-2 Minimum Nailing for Wood Structural Panel Applications

Application	Recommended Nail Size & Type	Nail Spacing (in.)	
		Panel Edges	Intermediate Supports
Single Floor–Glue-nailed installation[5]	**Ring- or screw-shank**		
16, 20, 24 oc, 3/4-in. thick or less	6d[1]	12	12
24 oc, 7/8-in. or 1-in. thick	8d[1]	6	12
32, 48 oc, 32-in. span (c-c)	8d[1]	6	12
48 oc, 48-in. span (c-c)	8d[2]	6	6
Single Floor–Nailed-only installation[5]	**Ring- or screw-shank**		
16, 20, 24 oc, 3/4-in. thick or less	6d	6	12
24 oc, 7/8-in. or 1-in. thick	8d	6	12
32, 48 oc, 32-in. span	8d[2]	6	12
48 oc, 48-in. span	8d[2]	6	6
Sheathing–Subflooring[3]	**Common smooth, ring- or screw-shank**		
7/16-in. to 1/2-in. thick	6d	6	12
7/8-in. thick or less	8d	6	12
Thicker panels	10d	6	6
Sheathing–Wall sheathing	**Common smooth, ring- or screw-shank or galvanized box[3]**		
1/2-in. thick or less	6d	6	12
Over 1/2-in. thick	8d	6	12
Sheathing–Roof sheathing	**Common smooth, ring- or screw-shank[3]**		
5/16-in. to 1-in. thick	8d	6	12[4]
Thicker panels	8d ring- or screw-shank or 10d common smooth	6	12[4]

1. 8d common nails may be substituted if ring- or screw-shank nails are not available.
2. 10d ring-shank, screw-shank, or common nails may be substituted if supports are well seasoned.
3. Other code-approved fasteners may be used.
4. For spans 48 in. or greater, space nails 6 in. at all supports.
5. Where required by the authority having jurisdiction, increased nailing schedules may be required.

M10:
MECHANICAL
CONNECTIONS

M10.1	**General**	**70**
M10.2	**Reference Design Values**	**71**
M10.3	**Design Adjustment Factors**	**71**
M10.4	**Typical Connection Details**	**72**
M10.5	**Pre-Engineered Metal Connectors**	**80**

10

M10.1 General

This Chapter covers design of connections between wood members using metal fasteners. Several common connection types are outlined below.

Dowel-Type (Nails, Bolts, Screws, Pins)

These connectors rely on metal-to-wood bearing for transfer of lateral loads and on friction or mechanical interfaces for transfer of axial (withdrawal) loads. They are commonly available in a wide range of diameters and lengths. More information is provided in Chapter M11.

Split Rings and Shear Plates

These connectors rely on their geometry to provide larger metal-to-wood bearing areas per connector. Both are installed into precut grooves or daps in the members. More information is provided in Chapter M12.

Timber Rivets

Timber rivets are a dowel-type connection, however, because the ultimate load capacity of such connections are limited by rivet bending and localized crushing of wood at the rivets or by the tension or shear strength of the wood at the perimeter of the rivet group, a specific design procedure is required. Timber rivet design loads are based on the lower of the maximum rivet bending load and the maximum load based on wood strength. Chapter M13 contains more information on timber rivet design.

Structural Framing Connections

Structural framing connections provide a single-piece connection between two framing members. They generally consist of bent or welded steel, carrying load from the supported member (through direct bearing) into the supporting member (by hanger flange bearing, fastener shear, or a combination of the two). Structural framing connections are proprietary connectors and are discussed in more detail in M10.4.

Other Connectors

Just as the number of possible building geometries is limitless, so too is the number of possible connection geometries. In addition to providing custom fabrication of connectors to meet virtually any geometry that can be designed, metal connector manufacturers have several categories of connectors that do not fit the categories above, including:

- framing anchors
- holddown devices
- straps and ties

These connectors are also generally proprietary connectors. See the manufacturer's literature or M10.4 for more information regarding design.

Connections are designed so that no applicable capacity is exceeded under loads. Strength criteria include lateral or withdrawal capacity of the connection, and tension or shear in the metal components. Some types of connections also include compression perpendicular to grain as a design criteria.

Users should note that design of connections may also be controlled by serviceability limitations. These limitations are product specific and are discussed in specific product chapters.

Stresses in Members at Connections

Local stresses in connections using multiple fasteners can be evaluated in accordance with *NDS* Appendix E.

M10.2 Reference Design Values

Reference design values for mechanical connections are provided in various sources. The *NDS* contains reference design values for dowel-type connections such as nails, bolts, lag screws, wood screws, split rings, shear plates, drift bolts, drift pins, and timber rivets.

Pre-engineered metal connectors are proprietary and reference design values are provided in code evaluation reports. More information on their use is provided in M10.5.

Metal connector plates are proprietary connectors for trusses, and reference design values are provided in code evaluation reports.

Staples and many pneumatic fasteners are proprietary, and reference design values are provided in code evaluation reports.

M10.3 Design Adjustment Factors

To generate connection design capacities, reference design values for connections are multiplied by adjustment factors per *NDS* 10.3. Applicable adjustment factors for connections are defined in *NDS* Table 10.3.1. Table M10.3-1 shows the applicability of adjustment factors for connections in a slightly different format for the designer.

Table M10.3-1   Applicability of Adjustment Factors for Mechanical Connections[1]

	Allowable Stress Design	**Load and Resistance Factor Design**
Lateral Loads		
Dowel-Type Fasteners	$Z' = Z\ C_D\ C_M\ C_t\ C_g\ C_\Delta\ C_{eg}\ C_{di}\ C_{tn}$	$Z' = Z\ C_M\ C_t\ C_g\ C_\Delta\ C_{eg}\ C_{di}\ C_{tn}\ K_F\ \phi_z\ \lambda$
Split Ring and Shear Plate Connectors	$P' = P\ C_D\ C_M\ C_t\ C_g\ C_\Delta\ C_d\ C_{st}$	$P' = P\ C_M\ C_t\ C_g\ C_\Delta\ C_d\ C_{st}\ K_F\ \phi_z\ \lambda$
	$Q' = Q\ C_D\ C_M\ C_t\ C_g\ C_\Delta\ C_d$	$Q' = Q\ C_M\ C_t\ C_g\ C_\Delta\ C_d\ K_F\ \phi_z\ \lambda$
Timber Rivets	$P' = P\ C_D\ C_M\ C_t\ C_{st}$	$P' = P\ C_M\ C_t\ C_{st}\ K_F\ \phi_z\ \lambda$
	$Q' = Q\ C_D\ C_M\ C_t\ C_\Delta\ C_{st}$	$Q' = Q\ C_M\ C_t\ C_\Delta\ C_{st}\ K_F\ \phi_z\ \lambda$
Metal Plate Connectors	$Z' = Z\ C_D\ C_M\ C_t$	$Z' = Z\ C_M\ C_t\ K_F\ \phi_z\ \lambda$
Spike Grids	$Z' = Z\ C_D\ C_M\ C_t\ C_\Delta$	$Z' = Z\ C_M\ C_t\ C_\Delta\ K_F\ \phi_z\ \lambda$
Withdrawal Loads		
Nails, Spikes, Lag Screws, Wood Screws, and Drift Pins	$W' = W\ C_D\ C_M\ C_t\ C_{eg}\ C_{tn}$	$Z' = Z\ C_M\ C_t\ C_{eg}\ C_{tn}\ K_F\ \phi_z\ \lambda$

1. See *NDS* Table 10.3.1 footnotes for additional guidance on application of adjustment factors for mechanical connections.

The following connection product chapters contain examples of the application of adjustment factors to reference design values:

Chapter M11 – dowel-type fasteners,
Chapter M12 – split ring and shear plate connectors,
Chapter M13 – timber rivets.

M10.4 Typical Connection Details

General Concepts of Well-Designed Connections

Connections must obviously provide the structural strength necessary to transfer loads. Well-designed connections hold the wood members in such a manner that shrinkage/swelling cycles do not induce splitting across the grain. Well-designed connections also minimize regions that might collect moisture – providing adequate clearance for air movement to keep the wood dry. Finally, well-designed connections minimize the potential for tension perpendicular to grain stresses – either under design conditions or under unusual loading conditions.

The following connection details (courtesy of the Canadian Wood Council) are organized into nine groups:

1. Beam to concrete or masonry wall connections
2. Beam to column connections
3. Column to base connections
4. Beam to beam connections
5. Cantilever beam connections
6. Arch peak connections
7. Arch base to support
8. Moment splice
9. Problem connections

Many of the detail groups begin with a brief discussion of the design challenges pertinent to the specific type of connection. Focusing on the key design concepts of a broad class of connections often leads to insights regarding a specific detail of interest.

Group 1. Beam to Concrete or Masonry Wall Connections

Design concepts. Concrete is porous and "wicks" moisture. Good detailing never permits wood to be in direct contact with concrete.

1. Beam on shelf in wall. The bearing plate distributes the load and keeps the beam from direct contact with the concrete. Steel angles provide uplift resistance and can also provide some lateral resistance. The end of the beam should not be in direct contact with the concrete.

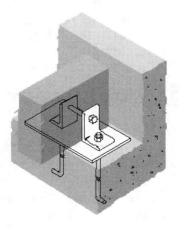

2. Similar to detail 1 with a steel bearing plate only under the beam.

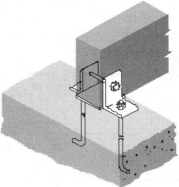

3. Similar to detail 1 with slotted holes to accommodate slight lateral movement of the beam under load. This detail is more commonly used when the beam is sloped, rather than flat.

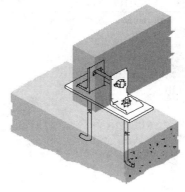

Group 2. Beam to Column Connections

Design concepts. All connections in the group must hold the beam in place on top of the column. Shear transfer is reasonably easy to achieve. Some connections must also resist some beam uplift. Finally, for cases in which the beam is spliced, rather than continuous over the column, transfer of forces across the splice may be required.

. Simple steel dowel for shear transfer.

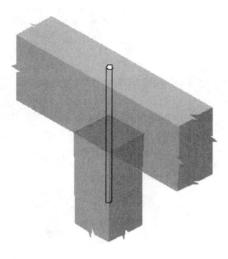

. Concealed connection in which a steel plate is inserted nto a kerf in both the beam and the column. Transverse ins or bolts complete the connection.

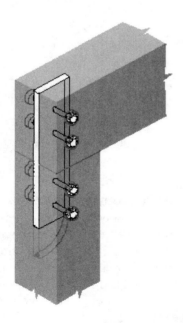

6. Custom welded column caps can be designed to transfer shear, uplift, and splice forces. Note design variations to provide sufficient bearing area for each of the beams and differing plate widths to accommodate differences between the column and the beam widths.

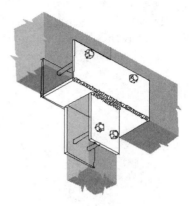

7. Combinations of steel angles and straps, bolted and screwed, to transfer forces.

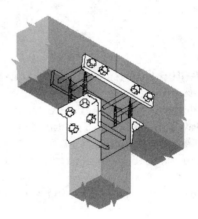

8. A very common connection – beam seat welded to the top of a steel column.

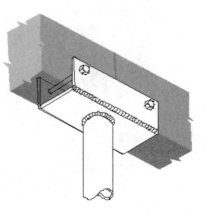

9. When both beams and columns are continuous and the connection must remain in-plane, either the beam or the column must be spliced at the connection. In this detail the column continuity is maintained. Optional shear plates may be used to transfer higher loads. Note that, unless the bolt heads are completely recessed into the back of the bracket, the beam end will likely require slotting. In a building with many bays, it may be difficult to maintain dimensions in the beam direction when using this connection.

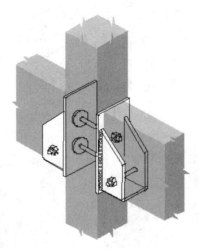

Group 3. Column to Base Connections

Design concepts. Since this is the bottom of the structure, it is conceivable that moisture from some source might run down the column. Experience has shown that base plate details in which a steel "shoe" is present can collect moisture that leads to decay in the column.

10. Similar to detail 4, with a bearing plate added.

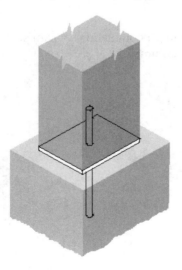

11A. Similar to details 1 and 2.

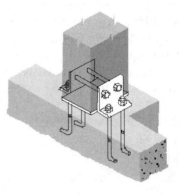

11B. Alternate to detail 11A.

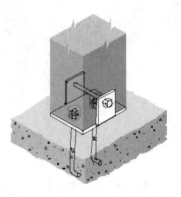

12. Similar to detail 3.

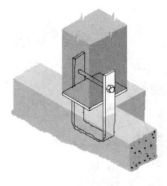

Group 4. Beam to Beam Connections

Design concepts. Many variations of this type of connection are possible. When all members are flat and their tops are flush, the connection is fairly straightforward. Slopes and skews require special attention to fabrication dimensions – well-designed connections provide adequate clearance to insert bolts or other connectors and also provide room to grip and tighten with a wrench. Especially for sloped members, special attention is required to visualize the stresses induced as the members deflect under load – some connections will induce large perpendicular to grain stresses in this mode.

3. Bucket-style welded bracket at a "cross" junction. The top of the support beam is sometimes dapped to accommodate the thickness of the steel.

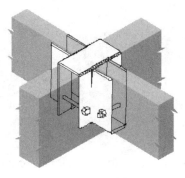

4. Face-mounted hangers are commonly used in beam to beam connections. In a "cross" junction special attention is required to fastener penetration length into the carrying beam (to avoid interference from other side).

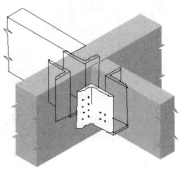

5. Deep members may be supported by fairly shallow hangers – in this case, through-bolted with shear plates. Clip angles are used to prevent rotation of the top of the suspended beam. Note that the clip angles are not connected to the suspended beam – doing so would restrain a deep beam from its natural across-the-grain shrinking and swelling cycles and would lead to splits.

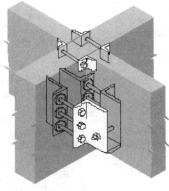

6. Concealed connections similar to detail 5. The suspended beam may be dapped on the bottom for a flush connection. The pin may be slightly narrower than the suspended beam, permitting plugging of the holes after the pin is installed. Note that the kerf in the suspended beam must accommodate not only the width of the steel plate, but also the increased width at the fillet welds.

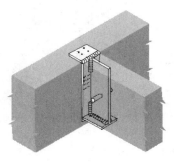

17. Similar to detail 13, with somewhat lower load capacity.

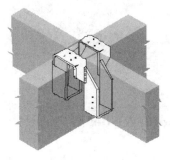

18. Clip angle to connect crossing beam.

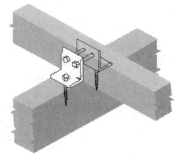

19. Special detail to connect the ridge purlin to sloped members or to the peak of arch members.

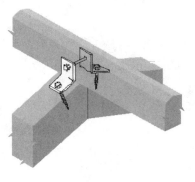

20A. Similar to detail 19, but with the segments of the ridge purlin set flush with the other framing.

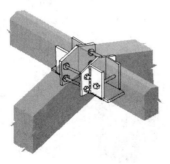

20B. Alternate to detail 20A.

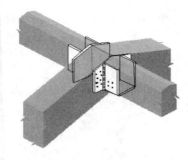

Group 5. Cantilever Beam Connections

21. Hinge connector transfers load without need to slope cut member ends. Beams are often dapped top and bottom for a flush fit.

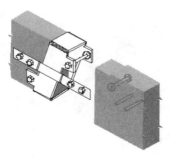

Group 6. Arch Peak Connections

22. Steep arches connected with a rod and shear plates.

23. Similar to detail 22, with added shear plate.

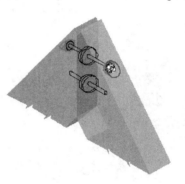

24. Similar to detail 22 for low slope arches. Side plate replace the threaded rod.

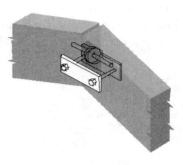

Group 7. Arch Base to Support

Design concepts. Arches transmit thrust into the sup porting structure. The foundation may be designed to resis this thrust or tie rods may be used. The base detail shoul be designed to accommodate the amount of rotation an ticipated in the arch base under various loading condition Elastomeric bearing pads can assist somewhat in distrib uting stresses. As noted earlier, the connection should b designed to minimize any perpendicular to grain stresse during the deformation of the structure under load.

25. Welded shoe transmits thrust from arch to suppor Note that inside edge of shoe is left open to prevent co lection of moisture.

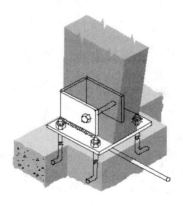

26. Arch base fastened directly to a steel tie beam in a shoe-type connection.

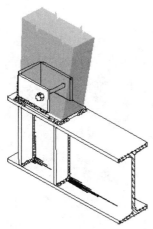

27. Similar to detail 25. This more rigid connection is suitable for spans where arch rotation at the base is small enough to not require the rotational movement permitted in detail 25. Note that, although the shoe is "boxed" a weep slot is provided at the inside face.

28. For very long spans or other cases where large rotations must be accommodated, a true hinge connection may be required.

Group 8. Moment Splice

Design concepts. Moment splices must transmit axial tension, axial compression, and shear. They must serve these functions in an area of the structure where structural movement may be significant – thus, they must not introduce cross-grain forces if they are to function properly.

29. Separate pieces of steel each provide a specific function. Top and bottom plate transfer axial force, pressure plates transfer direct thrust, and shear plates transmit shear.

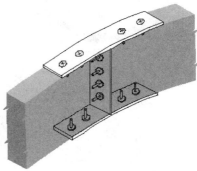

30. Similar to detail 29. Connectors on side faces may be easier to install, but forces are higher because moment arm between steel straps is less than in detail 29.

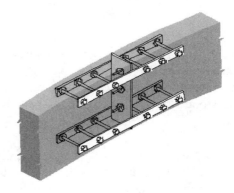

Group 9. Problem Connections

Hidden column base. It is sometimes preferable architecturally to conceal the connection at the base of the column. In any case it is crucial to detail this connection to minimize decay potential.

31A. Similar to detail 11, but with floor slab poured over the top of the connection. THIS WILL CAUSE DECAY AND IS NOT A RECOMMENDED DETAIL!

31B. Alternate to detail 31A.

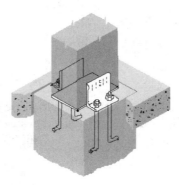

Full-depth side plates. It is sometimes easier to fabricate connections for deep beams from large steel plates rather than having to keep track of more pieces. Lack of attention to wood's dimensional changes as it "breathes" may lead to splits.

32A. Full-depth side plates may appear to be a good connection option. Unfortunately, the side plates will remain fastened while the wood shrinks over the first heating season. Since it is restrained by the side plates, the beam may split. THIS DETAIL IS NOT RECOMMENDED!

32B. As an alternative to detail 32A, smaller plates will transmit forces, but they do not restrain the wood from its natural movements.

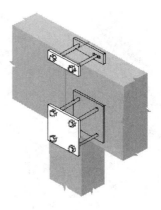

Notched beam bearing. Depth limitations sometimes cause detailing difficulties at the beam supports. A simple solution is to notch the beam at the bearing. This induces large tension perpendicular to grain stresses and leads to splitting of the beam at the root of the notch.

33A. Notching a beam at its bearing may cause splits. THIS DETAIL IS NOT RECOMMENDED!

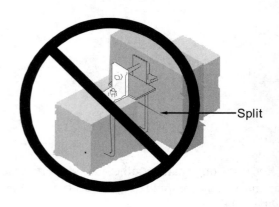

33B. Alternate to detail 33A.

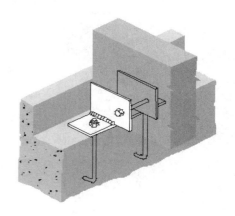

34A. This sloped bearing with a beam that is not fully supported may also split under load. THIS DETAIL IS NOT RECOMMENDED!

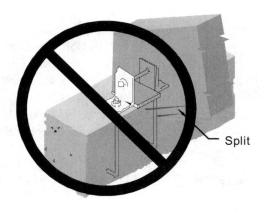

34B. Alternate to detail 34A.

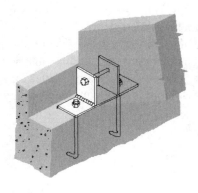

Hanging to underside of beam. Sometimes it is advantageous to hang a load from the underside of a beam. This is acceptable as long as the hanger is fastened to the upper half of the beam. Fastening to the lower half of the beam may induce splits.

35A. Connecting a hanger to the lower half of a beam that pulls downward may cause splits. THIS DETAIL IS NOT RECOMMENDED!

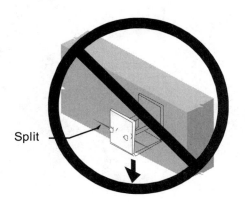

35B. As an alternative to detail 35A, the plates may be extended and the connection made to the upper half of the beam.

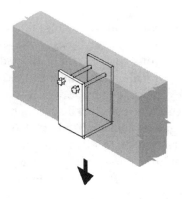

Hanger to side of beam. See full-depth side plates discussion.

36A. Deep beam hangers that have fasteners installed in the side plates toward the top of the supported beam may promote splits at the fastener group should the wood member shrink and lift from the bottom of the beam hanger because of the support provided by the fastener group. THIS DETAIL IS NOT RECOMMENDED!

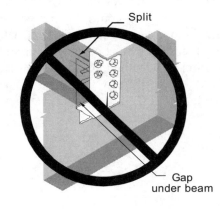

36B. Alternate to detail 36A.

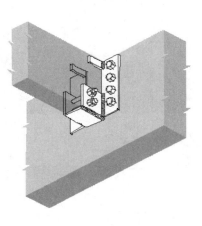

10

M10: MECHANICAL CONNECTIONS

M10.5 Pre-Engineered Metal Connectors

Product Information

Pre-engineered metal connectors for wood construction are commonly used in all types of wood construction. There are numerous reasons for their widespread use. Connectors often make wood members easier and faster to install. They increase the safety of wood construction, not only from normal loads, but also from natural disasters such as earthquakes and high winds. Connectors make wood structures easier to design by providing simpler connections of known load capacity. They also allow for the use of more cost-effective engineered wood members by providing the higher capacity connections often required by the use of such members. In certain locations, model building codes specifically require connectors or hangers.

Metal connectors are usually manufactured by stamping sheets or strips of steel, although some heavy hangers are welded together. Different thicknesses and grades of steel are used, depending on the required capacity of the connector.

Some metal connectors are produced as proprietary products which are covered by evaluation reports from one or all of the model building codes. Such reports should be consulted for current design information, while the manufacturer's literature can be consulted for additional design information and detailed installation instructions.

Common Uses

Pre-engineered metal connectors for wood construction are used throughout the world. Connectors are used to resist vertical dead, live, and snow loads; uplift loads from wind; and lateral loads from ground motion or wind. Almost any type of wood member may be fastened to another using a connector. Connectors may also be used to fasten wood to other materials, such as concrete, masonry, or steel.

Availability

Connectors are manufactured in varying load capacities, sizes, and configurations to fit a wide range of applications. A variety of connectors are widely available through lumber suppliers.

Because of the wide variety of available connectors, a generic design document such as this must be limited in its scope for simplicity's sake. Design values for specific connectors are available from the connector manufacturer.

ypes of Connectors

There are many different types of connectors, due to he many different applications in which connectors may be used. The following sections list the most common ypes of connectors.

ace Mount Joist Hangers

Face mount joist hangers install on the face of the supporting member, and rely on the shear capacity of the hails to provide holding power. Although referred to as oist hangers, these connectors may support other horizontal members subject to vertical loads, such as beams or purlins.

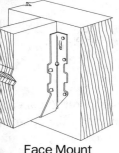

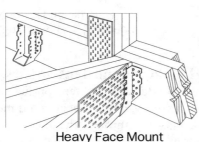

Face Mount
Joist Hanger

Heavy Face Mount
Joist Hanger

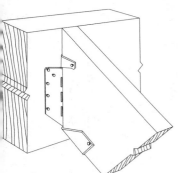

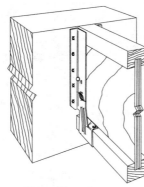

Slope- and Skew-
Adjustable Joist Hanger

Face Mount
I-Joist Hanger*

Top Flange Joist Hangers

Top flange joist, beam, and purlin hangers rely on bearing of the top flange onto the top of the supporting member, along with the shear capacity of any fasteners that are present in the face. Although referred to as joist hangers, these connectors may support other horizontal members subject to vertical loads, such as beams or purlins.

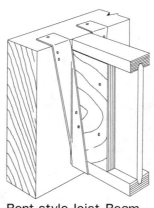

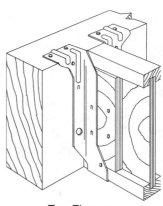

Bent-style Joist, Beam,
and Purlin Hanger*

Top Flange
I-joist Hanger*

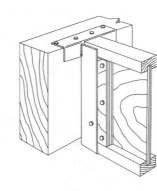

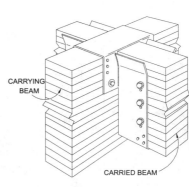

CARRYING
BEAM

CARRIED BEAM

Welded-type Purlin,
Beam, and Joist Hanger*

Heavy Welded Beam
Hanger (Saddle Type)

* Hangers should be capable of providing lateral support to the top flange of the joist. This is usually accomplished by a hanger flange that extends the full depth of the joist. At a minimum, hanger support should extend to at least mid-height of a joist used with web stiffeners. Some connector manufacturers have developed hangers specifically for use with wood I-joists that provide full lateral support without the use of web stiffeners.

Adjustable Style Hanger

Adjustable style joist and truss hangers have straps which can be either fastened to the face of a supporting member, similar to a face mount hanger, or wrapped over the top of a supporting member, similar to a top flange hanger.

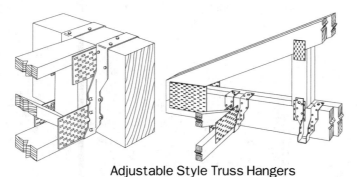

Adjustable Style Truss Hangers

Seismic and Hurricane Ties

Seismic and hurricane ties are typically used to connect two members that are oriented 90° from each other. These ties resist forces through the shear capacity of the nails in the members. These connectors may provide resistance in three dimensions.

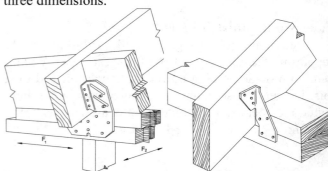

Seismic and Hurricane Ties
Connecting Roof Framing to Top Plates

Flat Straps

Flat straps rely on the shear capacity of the nails in the wood members to transfer load.

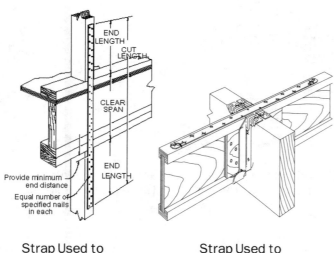

Strap Used to
Transfer Uplift Forces

Strap Used to
Transfer Lateral Forces

Holddowns and Tension Ties

Holddowns and tension ties usually bolt to concrete or masonry, and connect wood members to the concrete or masonry through the shear resistance of either nails, screws, or bolts. They may also be used to connect two wood members together.

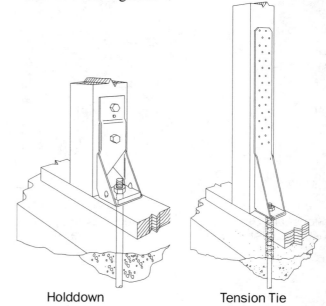

Holddown

Tension Tie

Embedded Type Anchors

Embedded anchors connect a wood member to concrete or masonry. One end of the connector embeds in the concrete or masonry, and the other end connects to the wood through the shear resistance of the nails or bolts.

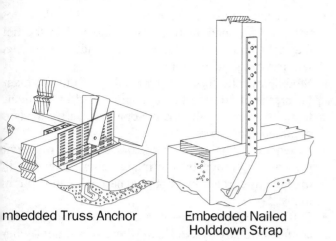

Embedded Truss Anchor   Embedded Nailed Holddown Strap

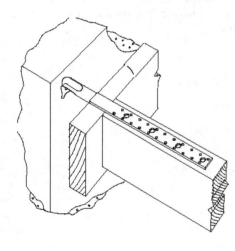

Purlin Anchor

Product Selection

Proper choice of connectors is required to optimize performance and economics. The selection of a connector will depend on several variables. These include the following:

- capacity required
- size and type of members being connected
- species of wood being connected
- slope and/or skew of member
- connector type preference

- type of fasteners to be used
- corrosion resistance desired
- appearance desired

Once the listed information is known, proper selection is facilitated through the use of manufacturer's literature, code evaluation reports, and software available from connector manufacturers.

This Manual provides guidance for specifying pre-engineered metal connectors to satisfy specific design criteria for a given application.

Connection Details

Connections, including pre-engineered metal connections, must provide the structural strength necessary to transfer loads. Well-designed connections hold wood members in such a manner that shrinkage/swelling cycles do not induce splitting across the grain. Well-designed connections also minimize collection of moisture – providing adequate clearance for air movement to keep the wood dry. Finally, well-designed connections minimize the potential for tension perpendicular to grain stresses – either under design conditions or under unusual loading conditions. Section M10.4 contains general concepts of well designed connections, including over 40 details showing acceptable and unacceptable practice.

Other Considerations

With proper selection and installation, structural connectors will perform as they were designed. However, proper selection and installation involves a variety of items that both the designer and the installer must consider including the general topics of: the wood members being connected; the fasteners used; and the connectors themselves. These items are discussed in the following sections. This Manual does not purport to address these topics in an all-inclusive manner – it is merely an attempt to alert designers to the importance of selection and installation details for achieving the published capacity of the connector.

Wood Members

The wood members being connected have an impact on the capacity of the connection. The following are important items regarding the wood members themselves:

- The species of wood must be the same as that for which the connector was rated by the manufacturer. Manufacturers test and publish allowable design values only for certain species of wood. For other species, consult with the connector manufacturer.

- The wood must not split when the fastener is installed. A fastener that splits the wood will not take the design load. If wood tends to split, consider pre-boring holes using a diameter not exceeding 3/4 of the nail diameter. Pre-boring requirements for screws and bolts are provided in the *NDS*.
- Wood can shrink and expand as it loses and gains moisture. Most connectors are manufactured to fit common dry lumber dimensions. Other dimensions may be available from the manufacturer.
- Where built-up lumber (multiple members) is installed in a connector, the members must be fastened together prior to installation of the connector so that the members act as a single unit.
- The dimensions of the supporting member must be sufficient to receive the specified fasteners. Most connectors are rated based on full penetration of all specified fasteners. Refer to the connector manufacturer for other situations.
- Bearing capacity of the joist or beam should also be evaluated to ensure adequate capacity.

Fasteners

Most wood connectors rely on the fasteners to transfer the load from one member to the other. Therefore, the choice and installation of the fasteners is critical to the performance of the connector.

The following are important items regarding the fasteners used in the connector:

- All fasteners specified by the manufacturer must be installed to achieve the published value.
- The size of fastener specified by the manufacturer must be installed. Most manufacturers specify common nails, unless otherwise noted.
- The fastener must have at least the same corrosion resistance as the connector.
- Bolts must generally be structural quality bolts, equal to or better than ANSI/ASME Standard B18.2.1.
- Bolt holes must be a minimum of 1/32" and a maximum of 1/16" larger than the bolt diameter.
- Fasteners must be installed prior to loading the connection.
- Pneumatic or powder-actuated fasteners may deflect and injure the operator or others. Nail guns may be used to install connectors, provided the correct quantity and type of nails are properly installed in the manufacturer's nail holes. Guns with nail hole-locating mechanisms should be used. Follow the nail gun manufacturer's instructions and use the appropriate safety equipment.

Connectors

Finally, the condition of the connector itself is critical to how it will perform. The following are important items regarding the connector itself:

- Connectors may not be modified in the field unless noted by the manufacturer. Bending steel in the field may cause fractures at the bend line, and fractured steel will not carry the rated load.
- Modified connectors may be available from the manufacturer. Not all modifications are tested by all manufacturers. Contact the manufacturer to verify loads of modified connectors.
- In general, all holes in connectors should be filled with the nails specified by the manufacturer. Contact the manufacturer regarding optional nail holes and optional loads.
- Different environments can cause corrosion of steel connectors. Always evaluate the environment where the connector will be installed. Connectors are available with differing corrosion resistances. Contact the manufacturer for availability. Fasteners must be at least the same corrosion resistance as that chosen for the connector.

M11: DOWEL-TYPE FASTENERS

M11.1 General   86

M11.2 Reference Withdrawal Design Values 86

M11.3 Reference Lateral Design Values   86

M11.4 Combined Lateral and Withdrawal
       Loads   86

M11.5 Adjustment of Reference
       Design Values   87

M11.6 Multiple Fasteners   87

11

M11.1 General

This Chapter covers design of connections between wood members using metal dowel-type (nails, bolts, lag screws, wood screws, drift pins) fasteners.

These connectors rely on metal-to-wood bearing for transfer of lateral loads and on friction or mechanical interfaces for transfer of axial (withdrawal) loads. They are commonly available in a wide range of diameters and lengths.

M11.2 Reference Withdrawal Design Values

The basic design equation for dowel-type fasteners under withdrawal loads is:

$$W'p \geq R_W$$

where:

W' = adjusted withdrawal design value

R_W = axial (withdrawal) force

p = depth of fastener penetration into wood member

Reference withdrawal design values are tabulated in *NDS* Chapter 11.

M11.3 Reference Lateral Design Values

The basic equation for design of dowel-type fasteners under lateral load is:

$$Z' \geq R_Z$$

where:

Z' = adjusted lateral design value

R_Z = lateral force

Reference lateral design values are tabulated in *NDS* Chapter 11.

M11.4 Combined Lateral and Withdrawal Loads

Lag screws, wood screws, nails, and spikes resisting combined lateral and withdrawal loads shall be designed in accordance with *NDS* 11.4.

M11.5 Adjustment of Reference Design Values

Dowel-type connections must be designed by applying all applicable adjustment factors to the reference withdrawal design value or reference lateral design value for the connection. *NDS* Table 10.3-1 lists all applicable adjustment factors for dowel-type connectors. Table M11.3-1 shows the applicability of adjustment factors for dowel-type fasteners in a slightly different format for the designer.

Table M11.3-1 Applicability of Adjustment Factors for Dowel-Type Fasteners[1]

	Allowable Stress Design	Load and Resistance Factor Design
Lateral Loads		
Dowel-Type Fasteners	$Z' = Z\ C_D\ C_M\ C_t\ C_g\ C_\Delta\ C_{eg}\ C_{di}\ C_{tn}$	$Z' = Z\ C_M\ C_t\ C_g\ C_\Delta\ C_{eg}\ C_{di}\ C_{tn}\ K_F\ \phi_z\ \lambda$
Withdrawal Loads		
Nails, Spikes, Lag Screws, Wood Screws, and Drift Pins	$W' = W\ C_D\ C_M\ C_t\ C_{eg}\ C_{tn}$	$Z' = Z\ C_M\ C_t\ C_{eg}\ C_{tn}\ K_F\ \phi_z\ \lambda$

1. See *NDS* Table 10.3.1 footnotes for additional guidance on application of adjustment factors for dowel-type fasteners.

Example of a Dowel-Type Fastener Loaded Laterally

For a single dowel-type fastener installed in side grain perpendicular to the length of the wood member, meeting the end and edge distance and spacing requirements of *NDS* 11.5.1, used in a normal building environment (meeting the reference conditions of *NDS* 2.3 and 10.3), and not a nail or spike in diaphragm construction, the general equation for Z' reduces to:

for ASD:
$$Z' = Z\ C_D$$

for LRFD:
$$Z' = Z\ K_F\ \phi_z\ \lambda$$

Example of a Dowel-Type Fastener Loaded in Withdrawal

For a single dowel-type fastener installed in side grain perpendicular to the length of the wood member, used in a normal building environment (meeting the reference conditions of *NDS* 2.3 and 10.3), the general equation for W' reduces to:

for ASD:
$$W' = W\ C_D$$

for LRFD:
$$W' = W\ K_F\ \phi_z\ \lambda$$

Installation Requirements

To achieve stated design values, connectors must comply with installation requirements such as spacing of connectors, minimum edge and end distances, proper drilling of lead holes, and minimum fastener penetration.

M11.6 Multiple Fasteners

Local stresses in connections using multiple fasteners can be evaluated in accordance with *NDS* Appendix E.

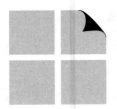

M12: SPLIT RING AND SHEAR PLATE CONNECTORS

M12.1 General   90

M12.2 Reference Design Values   90

**M12.3 Placement of Split Ring and
Shear Plate Connectors**   90

12

M12.1 General

This Chapter covers design for split rings and shear plates. These connectors rely on their geometry to provide larger metal-to-wood bearing areas per connector. Both are installed into precut grooves or daps in the members.

M12.2 Reference Design Values

Reference lateral design values (P, Q) are tabulated in the split ring and shear plate tables in *NDS* 12.2.

Design Adjustment Factors

Split ring and shear plate connections must be designed by applying all applicable adjustment factors to the reference lateral design value for the connection. *NDS* Table 10.3.1 provides all applicable adjustment factors for spl[it] ring and shear plate connectors. Table M12.2-1 shows th[e] applicability of adjustment factors for dowel-type fastener[s] in a slightly different format for the designer.

Table M12.2-1 Applicability of Adjustment Factors for Split Ring and Shear Plate Connectors[1]

	Allowable Stress Design		Load and Resistance Factor Design	
Split Ring and Shear Plate Connectors	$P' = P \, C_D \, C_M \, C_t \, C_g \, C_\Delta \, C_d \, C_{st}$		$P' = P \, C_M \, C_t \, C_g \, C_\Delta \, C_d \, C_{st} \, K_F \, \phi_z \, \lambda$	
	$Q' = Q \, C_D \, C_M \, C_t \, C_g \, C_\Delta \, C_d$		$Q' = Q \, C_M \, C_t \, C_g \, C_\Delta \, C_d \, K_F \, \phi_z \, \lambda$	

1. See *NDS* Table 10.3.1 footnotes for additional guidance on application of adjustment factors for split ring and shear plate connectors.

For a single split ring or shear plate connection installed in side grain perpendicular to the length of the wood members, meeting the end and edge distance and spacing requirements of *NDS* 12.3, used in a normal building environment (meeting the reference conditions of *NDS* 2.3 and 10.3), and meeting the penetration requirements of *NDS* 12.2.3, the general equations for P' and Q' reduce to:

for ASD:

$$P' = P \, C_D$$
$$Q' = Q \, C_D$$

for LRFD:

$$P' = P \, K_F \, \phi_z \, \lambda$$
$$Q' = Q \, K_F \, \phi_z \, \lambda$$

M12.3 Placement of Split Ring and Shear Plate Connectors

Installation Requirements

To achieve stated design values, connectors must comply with installation requirements such as spacing of connectors, minimum edge and end distances, proper dapping and grooving, drilling of lead holes, and minimum fastener penetration as specified in *NDS* 12.3.

M13: TIMBER RIVETS

M13.1 General   92

M13.2 Reference Design Values   92

M13.3 Placement of Timber Rivets   92

13

M13.1 General

This Chapter covers design for timber rivets. Timber rivets are hardened steel nails that are driven through pre-drilled holes in steel side plates (typically 1/4" thickness) to form an integrated connection where the plate and rivets work together to transfer load to the wood member.

M13.2 Reference Design Values

Reference wood capacity design values parallel to grain, P_w, are tabulated in the timber rivet Tables 13.2.1A through 13.2.1F in the *NDS*.

Reference design values perpendicular to grain are calculated per *NDS* 13.2.2.

for the connection. *NDS* Table 10.3-1 lists all applicabl adjustment factors for timber rivets. Table M13.2-1 show the applicability of adjustment factors for timber rivets i a slightly different format for the designer.

Design Adjustment Factors

Connections must be designed by applying all applicable adjustment factors to the reference lateral design value

Table M13.2-1   Applicability of Adjustment Factors for Timber Rivets[1]

	Allowable Stress Design	Load and Resistance Factor Design
Timber Rivets	$P' = P\ C_D\ C_M\ C_t\ C_{st}$	$P' = P\ C_M\ C_t\ C_{st}\ K_F\ \phi_z\ \lambda$
	$Q' = Q\ C_D\ C_M\ C_t\ C_\Delta$	$Q' = Q\ C_M\ C_t\ C_\Delta\ K_F\ \phi_z\ \lambda$

1. See *NDS* Table 10.3.1 footnotes for additional guidance on application of adjustment factors for timber rivets.

For a timber rivet connection installed in side grain perpendicular to the length of the wood members, with metal side plates 1/4" or greater, used in a normal building environment (meeting the reference conditions of *NDS* 2.3 and 10.3), and where wood capacity perpendicular to grain, Q_w, does not control, the general equations for P′ and Q′ reduce to:

for ASD:
$$P' = P\ C_D$$
$$Q' = Q\ C_D$$

for LRFD:
$$P' = P\ K_F\ \phi_z\ \lambda$$
$$Q' = Q\ K_F\ \phi_z\ \lambda$$

M13.3 Placement of Timber Rivets

Installation Requirements

To achieve stated design values, connectors must comply with installation requirements such as spacing of connectors, minimum edge and end distances per *NDS* 13.3; and drilling of lead holes, minimum fastener penetration, and other fabrication requirements per *NDS* 13.1.2.

M14: SHEAR WALLS AND DIAPHRAGMS

M14.1   General   94

M14.2   Design Principles   94

M14.3   Shear Walls   97

M14.4   Diaphragms   98

14

M14.1 General

This Chapter pertains to design of shear walls and diaphragms. These assemblies, which transfer lateral forces (wind and seismic) within the structure, are commonly designed using panel products fastened to framing members. The use of bracing systems to transfer these forces is not within the scope of this Chapter.

Shear wall/diaphragm shear capacity is tabulated i the *ANSI/AF&PA Special Design Provisions for Wind an Seismic (SDPWS) Supplement.*

M14.2 Design Principles

Drag Struts/Collectors

The load path for a box-type structure is from the diaphragm into the shear walls running parallel to the direction of the load (i.e., the diaphragm loads the shear walls that support it). Because the diaphragm acts like a long, deep beam, it loads each of the supporting shear walls evenly along the length of the walls. However, a wall typically contains windows and doors.

The traditional model used to analyze shear walls only recognizes full height wall segments as shear wall segments. This means that at locations with windows or doors, a structural element is needed to distribute diaphragm shear over the top of the opening and into the full height segments adjacent to it. This element is called a drag strut (see Figure M14.2-1).

In residential construction, the double top-plates existing in most stud walls will serve as a drag strut. It may be necessary to detail the double top plate such that no splices occur in critical zones. Or, it may be necessary to specify the use of a tension strap at butt joints to transfer these forces.

The maximum force seen by drag struts is generally equal to the diaphragm design shear in the direction of the shear wall multiplied by the distance between shear wall segments.

Drag struts are also used to tie together different parts of an irregularly shaped building.

To simplify design, irregularly shaped buildings (such as "L" or "T" shaped) are typically divided into simple rectangles. When the structure is "reassembled" after the individual designs have been completed, drag struts are used to provide the necessary continuity between these individual segments to insure that the building will act as a whole.

Figures M14.2-1, M14.2-2, and M14.2-3 and the accompanying generalized equations provide methods to calculate drag strut forces.

Figure M14.2-1 Shear Wall Drag Strut

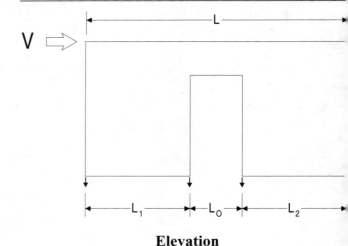

Elevation

Unit shear above opening $= \dfrac{V}{L} = v_a$

Unit shear below opening $= \dfrac{V}{L - L_0} = v_b$

Max. force in drag strut = greater of

$$\frac{\dfrac{V}{L} L_0 L_1}{(L - L_0)} = \frac{v_a L_0 L_1}{(L - L_0)}$$

or

$$\frac{\dfrac{V}{L} L_0 L_2}{(L - L_0)} = \frac{v_a L_0 L_2}{(L - L_0)}$$

Figure M14.2-2 Shear Wall Special Case Drag Strut

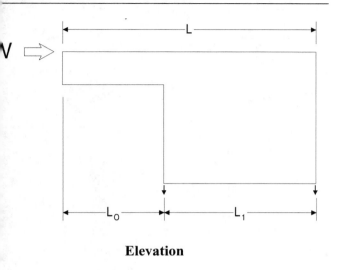

Elevation

Unit shear above opening $= \dfrac{V}{L} = v_a$

Unit shear below opening $= \dfrac{V}{L_1} = v_b$

Maximum force in drag strut $= L_o v_a$

Figure M14.2-3 Diaphragm Drag Strut
(Drag strut parallel to loads)

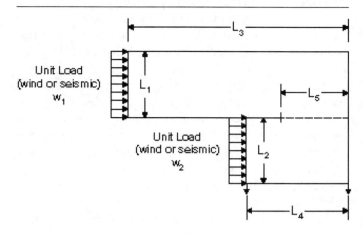

Unit shear along $L_3 = \dfrac{W_1 L_1}{2 L_3}$

Force in drag strut from L_3 structure $= \dfrac{W_1 L_1}{2 L_3}(L_5)$

Unit shear along $L_4 = \dfrac{W_2 L_2}{2 L_4}$

Force in drag strut from L_4 structure $= \dfrac{W_2 L_2}{2 L_4}(L_5)$

Maximum force in drag strut $= \dfrac{L_5}{2}\left(\dfrac{W_1 L_1}{L_3} + \dfrac{W_2 L_2}{L_4} \right)$

Chords

Diaphragms are assumed to act like long deep beams. This model assumes that shear forces are accommodated by the structural-use panel web of the "beam" and that moment forces are carried by the tension or compression forces in the flanges, or chords of the "beam." These chord forces are often assumed to be carried by the double top-plate of the supporting perimeter walls. Given the magnitude of forces involved in most light-frame wood construction projects, the double top-plate has sufficient capacity to resist tensile and compressive forces assuming adequate detailing at splice locations. However, offset wall lines and other factors sometimes make a continuous diaphragm chord impossible.

Because shear walls act as blocked, cantilevered diaphragms, they too develop chord forces and require chords. The chords in a shear wall are the double studs that are required at the end of each shear wall. Just as chords need to be continuous in a diaphragm, chords in a shear wall also need to maintain their continuity. This is accomplished by tension ties (holddowns) that are required at each end of each shear wall and between chords of stacked shear walls to provide overturning restraint. Figure M14.2-4 and the accompanying generalized equations provide a method to calculate chord forces.

Figure M14.2-4 Diaphragm Chord Forces

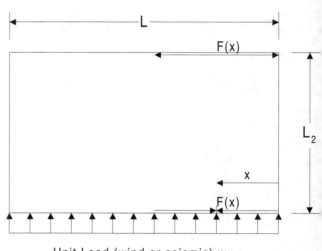

Unit Load (wind or seismic) w

Plan

$$\text{Diaphragm reaction} = \frac{Lw}{2}$$

$$\text{Diaphragm unit shear} = \frac{Lw}{2L_2}$$

$$\text{Diaphragm moment} = \frac{wL^2}{8}$$

$$\text{Maximum chord force} = \frac{wL^2}{8L_2}$$

$$\text{Chord force at point } x, \ F(x) = \frac{wLx}{2L_2} - \frac{wx^2}{2L_2}$$

M14.3 Shear Walls

Overturning

Overturning moments result from shear walls being loaded by horizontal forces. Overturning moments are resisted by force couples. The tension couple is typically achieved by a holddown. Figure M14.3-1 and the accompanying equations present a method for calculating overturning forces for a non-load-bearing wall. Figure M14.3-2 and the accompanying equations present a method for calculating overturning forces for a load-bearing wall. Overturning forces for load-bearing walls can utilize dead load as overturning restraint. To effectively resist uplift forces, holddown restraints are required to show very little slip relative to the chord (end post).

$$\text{Unit shear} = \frac{V}{L} = v$$

$$\text{Overturning force} = \text{chord force} = \frac{Vh}{L}$$

Figure M14.3-1 Overturning Forces
(no dead load)

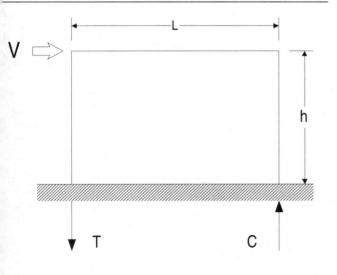

Elevation

Figure M14.3-2 Overturning Forces
(with dead load)

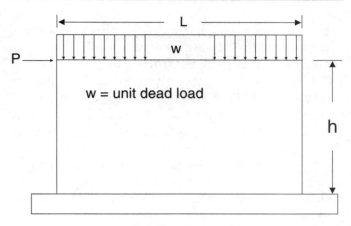

w = unit dead load

Elevation

$$\text{Overturning moment} = Ph$$

$$\text{Dead load restraining moment*} = \frac{wL^2}{2}$$

$$\text{Net overturning moment} = Ph - \frac{wL^2}{2}$$

$$\text{Net overturning force} - \text{chord force} = \frac{Ph - \frac{wL^2}{2}}{L} = \frac{Ph}{L} - \frac{wL}{2}$$

* See building code for applicable reduction to the dead load restraining moment to insure an appropriate load factor for overturning.

14

M14.4 Diaphragms

Subdiaphragms

The subdiaphragm (also known as the mini-diaphragm) concept has been recognized and extensively used to provide a method of meeting wall attachment and continuous cross-tie code requirements while minimizing the number and length of ties required to achieve continuity between chords. A formal definition of a subdiaphragm can be found in *SDPWS*, "SUBDIAPHRAGM portion of a larger wood diaphragm designed to anchor and transfer local forces to primary diaphragm struts and the main diaphragm."

In practice, the subdiaphragm approach is used to concentrate and transfer local lateral forces to the main structural members that support vertical loads. The subdiaphragm approach is often an economical solution to code required cross-ties for the following reasons:

- Main structural members are already present
- Main structural members generally span the full length and width of the buildings with few connectors.
- Main structural members are large enough to easily accommodate loads.
- Main structural members are large enough to allow "room" for requisite connections.

Each subdiaphragm must meet all applicable diaphragm requirements provided in the applicable building code. As such, each subdiaphragm must have chords, continuous tension ties, and sufficient sheathing thickness and attachment to transfer shear stresses generated within the diaphragm sheathing by the subdiaphragm. In addition, building codes may contain aspect ratios that are specific to subdiaphragms.

The subdiaphragm is actually the same structure as the main roof diaphragm, thus the subdiaphragm utilizes the same roof sheathing to transfer shear stresses as the main diaphragm. As such, sheathing nailing and thickness requirements of the roof diaphragm may not be sufficient for the subdiaphragm requirements. In this case, the subdiaphragm requirements would control and dictate roof sheathing and fastening requirements in subdiaphragm locations. Fortunately, the portion of the main diaphragm that is utilized as a subdiaphragm is a choice left to the designer; thus dimensions of the subdiaphragm can be chosen to minimize potential discontinuities in sheathing thicknesses or nail schedules. Similarly, the roof diaphragm requirements may be more stringent than those for the subdiaphragm.

M15: SPECIAL LOADING CONDITIONS

M15.1 Lateral Distribution of
       Concentrated Loads   100

M15.2 Spaced Columns   100

M15.3 Built-Up Columns   100

M15.4 Wood Columns with Side Loads
       and Eccentricity   100

15

M15.1 Lateral Distribution of Concentrated Loads

M15.1.1 Lateral Distribution of a Concentrated Load for Moment

The lateral distribution factors for moment in *NDS* Table 15.1.1 are keyed to the nominal thickness of the flooring or decking involved (2" to 6" thick). Spacing of the stringers or beams is based on recommendations of the American Association of State Highway and Transportation Officials.

Lateral distribution factors determined in accordance with *NDS* Table 15.1.1 can be used for any type of fixed or moving concentrated load.

M15.1.2 Lateral Distribution of a Concentrated Load for Shear

The lateral distribution factors for shear in *NDS* Tab 15.1.2 relate the lateral distribution of concentrated loa at the center of the beam or stringer span as determine under *NDS* 15.1.1, or by other means, to the distributio of load at the quarter points of the span. The quarter point are considered to be near the points of maximum shear i the stringers for timber bridge design.

M15.2 Spaced Columns

As used in the *NDS*, spaced columns refer to two or more individual members oriented with their longitudinal axis parallel, separated at the ends and in the middle portion of their length by blocking and joined at the ends by split ring or shear plate connectors capable of developing required shear resistance.

The end fixity developed by the connectors and en blocks increases the load-carrying capacity in compres sion parallel to grain of the individual members only i the direction perpendicular to their wide faces.

AF&PA's *Wood Structural Design Data (WSDL* provides load tables for spaced columns.

M15.3 Built-Up Columns

As with spaced columns, built-up columns obtain their efficiency by increasing the buckling resistance of individual laminations. The closer the laminations of a mechanically fastened built-up column deform together (the smaller the amount of slip occurring between laminations) under compressive load, the greater the relative capacity of that column compared to a simple solid column of the same slenderness ratio made with the same quality of material.

M15.4 Wood Columns with Side Loads and Eccentricity

The eccentric load design provisions of *NDS* 15.4.1 are not generally applied to columns supporting beam loads where the end of the beam bears on the entire cross section of the column. It is standard practice to consider such loads to be concentrically applied to the supporting column. This practice reflects the fact that the end fixity provided by the end of the column is ignored when the usual pinned end condition is assumed in column design. In applications where the end of the beam does not bear on the full cross section of the supporting column, or in

special critical loading cases, use of the eccentric colum loading provisions of *NDS* 15.4.1 may be considered ap propriate by the designer.

M16: FIRE DESIGN

M16.1 General   102

    Lumber   103

    Structural Glued Laminated
    Timber   119

    Poles and Piles   120

    Structural Composite Lumber   121

    Wood I-Joists   122

    Metal Plate Connected Wood
    Trusses   131

M16.2 Design Procedures for Exposed
      Wood Members   145

M16.3 Wood Connections   159

16

M16.1 General

This Chapter outlines fire considerations including design requirements and fire-rated assemblies for various wood products. Lumber, glued laminated timber, poles and piles, wood I-joists, structural composite lumber, and metal plate connected wood trusses are discussed.

Planning

As a first step, the authority having jurisdiction where a proposed building is to be constructed must be consulted for the requirements of the specific design project. This normally concerns the type of construction desired as well as allowable building areas and heights for each construction type.

Wood building construction is generally classified into types such as wood frame (Type V), noncombustible or fire-retardant-treated wood wall-wood joist (Type III), and heavy timber (Type IV). Type V construction is defined as having exterior walls, bearing walls, partitions, floors and roofs of wood stud and joist framing of 2" nominal dimension. These are divided into two subclasses that are either protected or unprotected construction. Protected construction calls for having load-bearing assemblies of 1-hour fire endurance.

Type III construction has exterior walls of noncombustible materials and roofs, floors, and interior walls and partitions of wood frame. As in Type V construction, these are divided into two subclasses that are either protected or unprotected.

Type IV construction includes exterior walls of noncombustible materials or fire-retardant-treated wood and columns, floors, roofs, and interior partitions of wood of a minimum size, as shown in Table M16.1-1.

In addition to having protected and unprotected subclasses for each building type, increases in floor area and height of the building are allowed when active fire protection, such as sprinkler protection systems, are included. For example, protected wood-frame business occupancies can be increased from three to four stories in height because of the presence of sprinklers. Also, the floor area may be further increased under some conditions. Additional information is available at www.awc.org.

Table M16.1-1 Minimum Sizes to Qualify as Heavy Timber Construction

Material	Minimum size (nominal size or thickness)
Roof decking: Lumber or structural-use panels	2 in. thickness 1-1/8 in. thickness
Floor decking: Lumber or flooring or structural-use panels	3 in. thickness 1 in. thickness 1/2 in. thickness
Roof framing:	4 by 6 in.
Floor framing:	6 by 10 in.
Columns:	8 by 8 in. (supporting floors) 6 by 8 in. (supporting roofs)

Building Code Requirements

For occupancies such as stores, apartments, offices, and other commercial and industrial uses, building codes commonly require floor/ceiling and wall assemblies to be fire-resistance rated in accordance with standard fire tests.

Depending on the application, wall assemblies may need to be rated either from one side or both sides. For specific exterior wall applications, the *International Building Code* (*IBC*) allows wood-frame, wood-sided walls to be tested for exposure to fire from the inside only. Rating for both interior and exterior exposure is only required when the wall has a fire separation distance of less than 5 feet. Code recognition of 1- and 2-hour wood-frame wall systems is also predicated on successful fire and hose stream testing in accordance with ASTM E119, *Standard Test Methods for Fire Tests of Building Construction Materials*.

Fire Tested Assemblies

Fire-rated wood-frame assemblies can be found in a number of sources including the *IBC*, Underwriters Laboratories (UL) *Fire Resistance Directory*, Intertek Testing Services' *Directory of Listed Products*, and the Gypsum Association's *Fire Resistance Design Manual*. The American Forest & Paper Association (AF&PA) and its members have tested a number of wood-frame fire-rated assemblies. Descriptions of these successfully tested assemblies are provided in Tables M16.1-2 through M16.1-5.

Updates

Additional tests are being conducted and the Tables will be updated periodically. AF&PA's *Design for Code Acceptance (DCA) No. 3, Fire Rated Wood Floor and Wall Assemblies* incorporates many of these assemblies and is available at www.awc.org.

Table M16.1-2 One-Hour Fire-Rated Load-Bearing Wood-Frame Wall Assemblies

		Assemblies Rated From Both Sides			
Studs	Insulation	Sheathing on Both Sides		Fasteners	Details
x4 @ 16" o.c.	3½" mineral wool batts	5/8" Type X Gypsum Wallboard (**H**)		2¼" #6 Type S drywall screws @ 12" o.c.	Figure M16.1-1
x6 @ 16" o.c.	(none)	5/8" Type X Gypsum Wallboard (**H**)		2¼" #6 Type S drywall screws @ 7" o.c.	Figure M16.1-2
x6 @ 16" o.c.	5½" mineral wool batts	5/8" Type X Gypsum Wallboard (**H**)		2¼" #6 Type S drywall screws @ 12" o.c.	Figure M16.1-3
x6 @ 16" o.c.	R-19 fiberglass insulation	5/8" Type X Gypsum Wallboard (**V**)		2¼" #6 Type S drywall screws @ 12" o.c.	Figure M16.1-4
		Assemblies Rated From One Side (Fire on Interior Only)			
Studs	Insulation	Sheathing		Fasteners	Details
x4 @ 16" o.c.	3½" mineral wool batts	I	5/8" Type X Gypsum Wallboard (**H**)	2¼" #6 Type S drywall screws @ 12" o.c.	Figure M16.1-5
		E	3/8" wood structural panels (**V**)	6d common nails @ 6" edges/12" field	
x4 @ 16" o.c.	4 mil polyethylene 3½" mineral wool batts	I	5/8" Type X Gypsum Wallboard (**V**)	6d cement coated box nails @ 7" o.c.	Figure M16.1-6
		E	½" fiberboard (**V**)	1½" roofing nails @ 3" edges/6" field	
			3/8" hardboard shiplapped panel siding	8d galv. nails @ 4" edges/8" field	
x6 @ 16" o.c.	5½" mineral wool batts	I	5/8" Type X Gypsum Wallboard (**H**)	2¼" #6 Type S drywall screws @ 12" o.c.	Figure M16.1-7
		E	7/16" wood structural panels (**V**)	6d common nails @ 6" edges/12" field	
x6 @ 16" o.c.	R-19 fiberglass insulation	I	5/8" Type X Gypsum Wallboard (**V**)	2¼" #6 Type S drywall screws @ 12" o.c.	Figure M16.1-8
		E	3/8" wood structural panels (**V**)	6d common nails @ 6" edges/12" field	

H- applied horizontally with vertical joints over studs; **I**- Interior sheathing; **V**- applied vertically with vertical joints over studs; **E**- Exterior sheathing

Table M16.1-3 Two-Hour Fire-Rated Load-Bearing Wood-Frame Wall Assemblies

		Assemblies Rated From Both Sides			
Studs	Insulation	Sheathing on Both Sides		Fasteners	Details
x4 @ 24" o.c.	5½" mineral wool batts	B	5/8" Type X Gypsum Wallboard (**H**)	2¼" #6 Type S drywall screws @ 24" o.c.	Figure M16.1-9
		F	5/8" Type X Gypsum Wallboard (**H**)	2¼" #6 Type S drywall screws @ 8" o.c.	

H- applied horizontally with vertical joints over studs; **B**- Base layer sheathing: **F**- Face layer sheathing

Table M16.1-4   One-Hour Fire-Rated Wood Floor/Ceiling Assemblies

Joists	Insulation	Furring		Ceiling Sheathing	Floor Sheathing	Details
2x10 @ 16" o.c.	none	Optional	F	5/8" Type X Gypsum Wallboard or ½" Type X Gypsum Wallboard	Nom. 1" wood flooring or 19/32" T&G plywood* underlayment (single floor); building paper, and Nom. 1" T&G boards or 15/32" plywood* subfloor	Figure M16.1-10
2x10 @ 16" o.c.	none	(none)	F	½" x 24" x 48" mineral acoustical ceiling panels (see grid details)	Nom. 19/32" T&G plywood* underlayment (single floor); building paper, and 15/32" plywood* subfloor	Figure M16.1-11
2x10 @ 16" o.c.	none	Resilient channels	F	5/8" Type X Gypsum Wallboard or ½" proprietary Type Gypsum Wallboard	Nom. 19/32" T&G plywood* underlayment (single floor) or 15/32" plywood* subfloor	Figure M16.1-12
2x10 @ 24" o.c.	none	Resilient channels	F	5/8" proprietary Type Gypsum Wallboard	Nom. 23/32" T&G plywood* underlayment (single floor) or 15/32" plywood* subfloor	Figure M16.1-13

F- Face layer sheathing; *Oriented strand board (OSB) panels are permitted for certain designs. Subfloors for certain designs may be nom. 7/16" OSB.

Table M16.1-5   Two-Hour Fire-Rated Wood Floor/Ceiling Assemblies

Joists	Insulation	Furring		Ceiling Sheathing	Floor Sheathing	Details
2x10 @ 16" o.c.	none	(none)	B	5/8" proprietary Type X Gypsum Wallboard	Nom. 1" wood flooring or 19/32" T&G plywood* underlayment (single floor); building paper, and Nom. 1" T&G boards or 15/32" plywood* subfloor	Figure M16.1-14
		Resilient channels	F	5/8" proprietary Type X Gypsum Wallboard		

B- Base layer sheathing (direct attached); F- Face layer sheathing; *Oriented strand board (OSB) panels are permitted for certain designs. Subfloors for certai designs may be nom. 7/16" OSB.

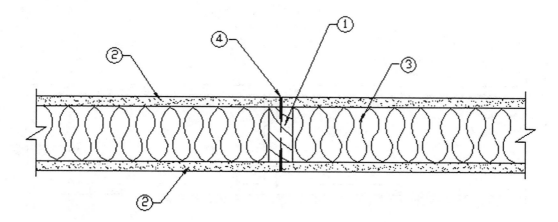

Framing: Nominal 2x4 wood studs, spaced 16 in. o.c., double top plates, single bottom plate.

Sheathing: 5/8 in. Type X gypsum wallboard, 4 ft. wide, applied horizontally, unblocked. Horizontal application of wallboard represents the direction of least fire resistance as opposed to vertical application.

Insulation: 3-1/2 in. thick mineral wool insulation.

Fasteners: 2-1/4 in. Type S drywall screws, spaced 12 in. o.c.

Joints and Fastener Heads: Wallboard joints covered with paper tape and joint compound, fastener heads covered with joint compound.

Tests conducted at the Fire Test Laboratory of National Gypsum Research Center

Test No: WP-1248 (Fire Endurance), March 29, 2000

WP-1246 (Hose Stream), March 9, 2000

Third-Party Witness: Intertek Testing Services

Report J20-06170.1

This assembly was tested at 100% design load, calculated in accordance with the *National Design Specification for Wood Construction*. The authority having jurisdiction should be consulted to assure acceptance of this report.

M16: FIRE DESIGN

16

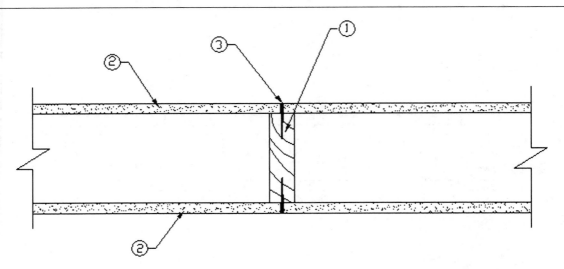

1. **Framing:** Nominal 2x6 wood studs, spaced 16 in. o.c., double top plates, single bottom plate.

2. **Sheathing:** 5/8 in. Type X gypsum wallboard, 4 ft. wide, applied horizontally, unblocked. Horizontal application of wallboard represents the direction of least fire resistance as opposed to vertical application.

3. **Fasteners:** 2-1/4 in. Type S drywall screws, spaced 7 in. o.c.

4. **Joints and Fastener Heads:** Wallboard joints covered with paper tape and joint compound, fastener heads covered with joint compound.

Tests conducted at the Fire Test Laboratory of Nationa Gypsum Research Center

Test No: WP-1232 (Fire Endurance), September 1(1999

WP-1234 (Hose Stream), September 27, 1999

Third-Party Witness: Intertek Testing Services

Report J99-22441.2

This assembly was tested at 100% design load, calculated in accordance with the *National Design Specification for Wood Construction*. The authority having jurisdiction should be consulted to assure acceptance of this report.

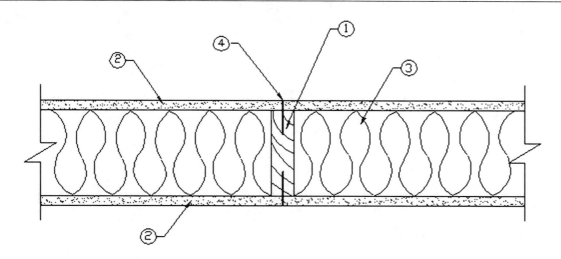

. **Framing:** Nominal 2x6 wood studs, spaced 16 in. o.c., double top plates, single bottom plate.

. **Sheathing:** 5/8 in. Type X gypsum wallboard, 4 ft. wide, applied horizontally, unblocked. Horizontal application of wallboard represents the direction of least fire resistance as opposed to vertical application.

. **Insulation:** 5-1/2 in. thick mineral wool insulation.

. **Fasteners:** 2-1/4 in. Type S drywall screws, spaced 12 in. o.c.

. **Joints and Fastener Heads:** Wallboard joints covered with paper tape and joint compound, fastener heads covered with joint compound.

Tests conducted at the Fire Test Laboratory of National Gypsum Research Center

Test No: WP-1231 (Fire Endurance), September 14, 1999

WP-1230 (Hose Stream), August 30, 1999

Third-Party Witness: Intertek Testing Services

Report J99-22441.1

This assembly was tested at 100% design load, calculated in accordance with the *National Design Specification for Wood Construction*. The authority having jurisdiction should be consulted to assure acceptance of this report.

Figure M16.1-4   One-Hour Fire-Resistive Wood Wall Assembly (WS6-1.4)
2x6 Wood Stud Wall - 100% Design Load - ASTM E119/NFPA 251

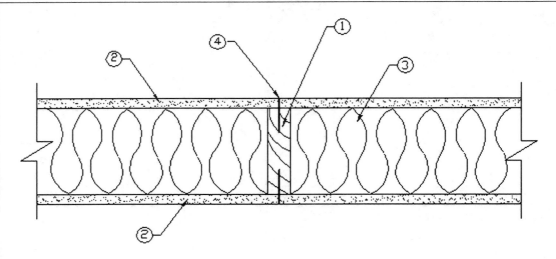

1. **Framing**: Nominal 2x6 wood studs, spaced 16 in. o.c., double top plates, single bottom plate

2. **Sheathing**: 5/8 in. Type X gypsum wallboard, 4 ft. wide, applied vertically. All panel edges backed by framing or blocking.

3. **Insulation**: R-19 fiberglass insulation.

4. **Fasteners**: 2-1/4 in. Type S drywall screws, spaced 12 in. o.c.

5. **Joints and Fastener Heads**: Wallboard joints covered with paper tape and joint compound, fastener heads covered with joint compound.

Tests conducted at NGC Testing Services

Test No: WP-1346 (Fire Endurance), August 22 2003

WP-1351 (Hose Stream), September 17, 2003

Third-Party Witness: NGC Testing Services

This assembly was tested at 100% design load, calculated in accordance with the *National Design Specification for Wood Construction*. The authority having jurisdiction should be consulted to assure acceptance of this report.

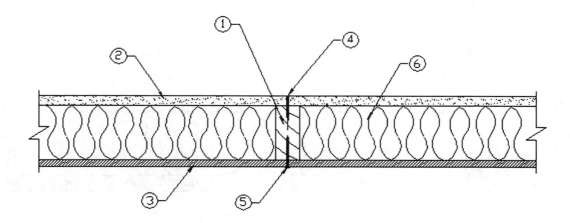

Framing: Nominal 2x4 wood studs, spaced 16 in. o.c., double top plates, single bottom plate.

Interior Sheathing: 5/8 in. Type X gypsum wallboard, 4 ft. wide, applied horizontally, unblocked. Horizontal application of wallboard represents the direction of least fire resistance as opposed to vertical application.

Exterior Sheathing: 3/8 in. wood structural panels (oriented strand board), applied vertically, horizontal joints blocked.

Gypsum Fasteners: 2-1/4 in. Type S drywall screws, spaced 12 in. o.c.

Panel Fasteners: 6d common nails (bright): 12 in. o.c. in the field, 6 in. o.c. panel edges.

Insulation: 3-1/2 in. thick mineral wool insulation.

Joints and Fastener Heads: Wallboard joints covered with paper tape and joint compound, fastener heads covered with joint compound.

Tests conducted at the Fire Test Laboratory of National Gypsum Research Center

Test No: WP-1261 (Fire Endurance & Hose Stream), November 1, 2000

Third-Party Witness: Intertek Testing Services

Report J20-006170.2

This assembly was tested at 100% design load, calculated in accordance with the *National Design Specification for Wood Construction*. The authority having jurisdiction should be consulted to assure acceptance of this report.

M16: FIRE DESIGN

16

Figure M16.1-6   One-Hour Fire-Resistive Wood Wall Assembly (WS4-1.3)
2x4 Wood Stud Wall - 78% Design Load - ASTM E119/NFPA 251

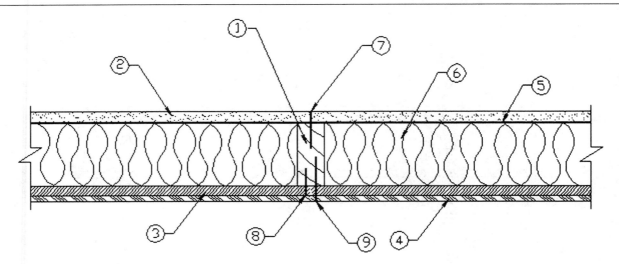

1. **Framing:** Nominal 2x4 wood studs, spaced 16 in. o.c., double top plates, single bottom plate.

2. **Interior Sheathing:** 5/8 in. Type X gypsum wallboard, 4 ft. wide, applied vertically, unblocked.

3. **Exterior Sheathing:** 1/2 in. fiberboard sheathing. *Alternate construction: minimum 1/2 in. lumber siding or 1/2 in. wood-based sheathing.*

4. **Exterior Siding:** 3/8 in. hardboard shiplap edge panel siding. *Alternate construction lumber or wood-based, vinyl, or aluminum siding.*

5. **Vapor Barrier:** 4-mil polyethylene sheeting.

6. **Insulation:** 3-1/2 in. thick mineral wool insulation.

7. **Gypsum Fasteners:** 6d cement coated box nails spaced 7 in. o.c.

8. **Fiberboard Fasteners:** 1-1/2 in. galvanized roofing nails: 6 in. o.c. in the field, 3 in. o.c. panel edges.

9. **Hardboard Fasteners:** 8d galvanized nails: 8 in. o.c. in the field, 4 in. o.c. panel edges.

10. **Joints and Fastener Heads:** Wallboard joints covered with paper tape and joint compound, fastener heads covered with joint compound.

Tests conducted at the Gold Bond Building Product Fire Testing Laboratory

Test No: WP-584 (Fire Endurance & Hos Stream), March 19, 1981

Third-Party Witness: Warnock Hersey Internationa Inc.

Report WHI-690-003

This assembly was tested at 78% design load using an ℓ_e/d of 33, calculated in accordance with the 1997 *National Design Specification® for Wood Construction*. The authority having jurisdiction should be consulted to assure acceptance of this report.

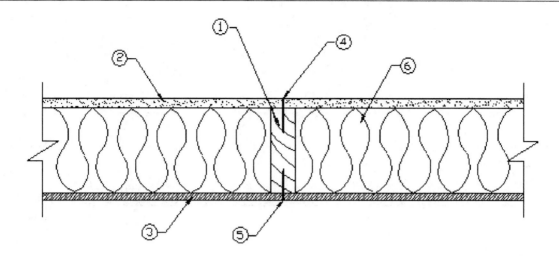

1. **Framing:** Nominal 2x6 wood studs, spaced 16 in. o.c., double top plates, single bottom plate.

2. **Interior Sheathing:** 5/8 in. Type X gypsum wallboard, 4 ft. wide, applied horizontally, unblocked. Horizontal application of wallboard represents the direction of least fire resistance as opposed to vertical application.

3. **Exterior Sheathing:** 7/16 in. wood structural panels (oriented strand board), applied vertically, horizontal joints blocked.

4. **Gypsum Fasteners:** 2-1/4 in. Type S drywall screws, spaced 12 in. o.c.

5. **Panel Fasteners:** 6d common nails (bright): 12 in. o.c. in the field, 6 in. o.c. panel edges.

6. **Insulation:** 5-1/2 in. thick mineral wool insulation.

7. **Joints and Fastener Heads:** Wallboard joints covered with paper tape and joint compound, fastener heads covered with joint compound.

Tests conducted at the Fire Test Laboratory of National Gypsum Research Center

Test No: WP-1244 (Fire Endurance & Hose Stream), February 25, 2000

Third-Party Witness: Intertek Testing Services

Report J99-27259.2

This assembly was tested at 100% design load, calculated in accordance with the *National Design Specification for Wood Construction*. The authority having jurisdiction should be consulted to assure acceptance of this report.

M16: FIRE DESIGN

16

Figure M16.1-8   One-Hour Fire-Resistive Wood Wall Assembly (WS6-1.5)
2x6 Wood Stud Wall - 100% Design Load - ASTM E119/NFPA 25

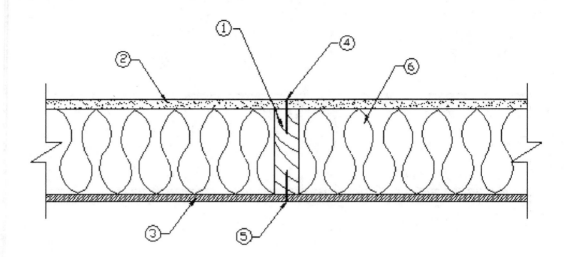

1. **Framing**: Nominal 2x6 wood studs, spaced 16 in. o.c., double top plates, single bottom plate.

2. **Interior Sheathing:** 5/8 in. Type X gypsum wallboard, 4 ft. wide, applied vertically. All panel edges backed by framing or blocking.

3. **Exterior Sheathing:** 3/8 in. wood structural panels (oriented strand board), applied vertically, horizontal joints blocked.

4. **Gypsum Fasteners:** 2-1/4 in. Type S drywall screws, spaced 7 in. o.c.

5. **Panel Fasteners:** 6d common nails (bright) - 12 in. o.c. in the field, 6 in. o.c. panel edges.

6. **Insulation:** R-19 fiberglass insulation.

7. **Joints and Fastener Heads:** Wallboard joints covered with paper tape and joint compound, fastener heads covered with joint compound.

Tests conducted at the NGC Testing Services

Test No: WP-1408 (Fire Endurance & Hose Stream August 13, 2004

Third-Party Witness: NGC Testing Services

This assembly was tested at 100% design load, calculated in accordance with the *National Design Specification for Wood Construction*. The authority having jurisdiction should be consulted to assure acceptance of this report.

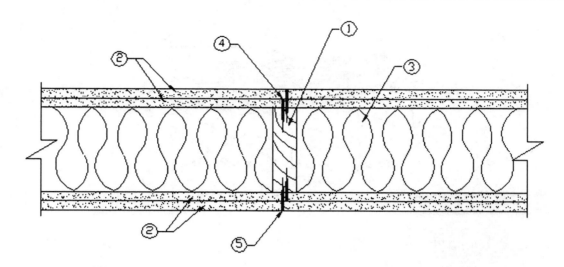

Framing: Nominal 2x6 wood studs, spaced 24 in. o.c., double top plates, single bottom plate.

Sheathing:

Base Layer: 5/8 in. Type X gypsum wallboard, 4 ft. wide, applied horizontally, unblocked.

Face Layer: 5/8 in. Type X gypsum wallboard, 4 ft. wide, applied horizontally, unblocked. Horizontal application of wallboard represents the direction of least fire resistance as opposed to vertical application.

Insulation: 5-1/2 in. thick mineral wool insulation.

Gypsum Fasteners: Base Layer: 2-1/4 in. Type S drywall screws, spaced 24 in. o.c.

Gypsum Fasteners: Face Layer: 2-1/4 in. Type S drywall screws, spaced 8 in. o.c.

Joints and Fastener Heads: Wallboard joints covered with paper tape and joint compound, fastener heads covered with joint compound.

Tests conducted at the Fire Test Laboratory of National Gypsum Research Center

Test No: WP-1262 (Fire Endurance), November 3, 2000

WP-1268 (Hose Stream), December 8, 2000

Third-Party Witness: Intertek Testing Services

Report J20-006170.3

This assembly was tested at 100% design load, calculated in accordance with the *National Design Specification for Wood Construction*. The authority having jurisdiction should be consulted to assure acceptance of this report.

M16: FIRE DESIGN

16

Figure M16.1-10   One-Hour Fire-Resistive Wood Floor/Ceiling Assembly

(2x10 Wood Joists 16" o.c. – Gypsum Directly Applied or on Optional Resilient Channels)

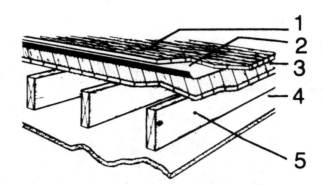

1. Nominal 1 in. wood flooring or 19/32 in. T&G plywood underlayment (single floor). Oriented strand board (OSB) panels are permitted for certain designs.

2. Building paper.

3. Nominal 1 in. T&G boards or 15/32 in. plywood subfloor. Subfloors for certain designs may be nominal 7/16 in. OSB.

4. 1/2 in. Type X gypsum wallboard (may be attached directly to joists or on resilient channels) or 5/8 in. Type X gypsum wallboard directly applied to joists.

5. 2x10 wood joists spaced 16 in. o.c.

Fire Tests:

1/2 in. Type X gypsum directly applied

    UL R1319-66, 11-9-64, Design L512;

    UL R3501-45, 5-27-65, Design L522;

    UL R2717-38, 6-10-65, Design L503;

    UL R3543-6, 11-10-65, Design L519;

    ULC Design M502

1/2 in. Type X gypsum on resilient channels

    UL R1319-65, 11-16-64, Design L514

5/8 in. Type X Gypsum directly applied

    UL R3501-5, 9, 7-15-52;

    UL R1319-2, 3, 6-5-52, Design L 501;

    ULC Design M500

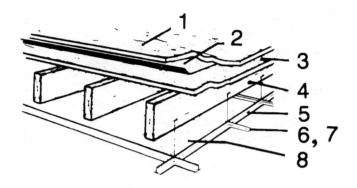

Nominal 19/32 in. T&G plywood underlayment (single floor). Oriented strand board (OSB) panels are permitted for certain designs.

Building paper.

15/32 in. plywood subfloor. Subfloors for certain designs may be nom. 7/16 in. OSB.

2x10 wood joists spaced 16 in. o.c.

T-bar grid ceiling system.

Main runners spaced 48 in. o.c.

Cross-tees spaced 24 in. o.c.

1/2 in. x 24 in. x 48 in. mineral acoustical ceiling panels installed with holddown clips.

Fire Tests:

UL L209

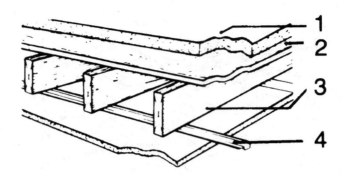

1. 1-1/2 in. lightweight concrete or minimum 3/4 in. proprietary gypsum concrete floor topping. Building paper may be optional and is not shown.

2. 15/32 in. plywood subfloor (subfloors for certain designs may be nominal 7/16 in. OSB) or nominal 19/32 in. T&G plywood underlayment (single floor). Oriented strand board (OSB) panels are permitted for certain designs.

3. 2x10 wood joists spaced 16 in. o.c.

4. 5/8 in. Type X Gypsum Wallboard or 1/2 in. proprietary Type X Gypsum Wallboard ceiling attached to resilient channels.

Fire Tests:

UL R1319-65, 11-16-64, Design L514 5/8 in. Type X gypsum

UL R6352, 4-21-71, Design L502 1/2 in. proprietary Type X gypsum

Figure M16.1-13 One-Hour Fire-Resistive Wood Floor/Ceiling Assembly
(2x10 Wood Joists 24" o.c. – Gypsum on Resilient Channels)

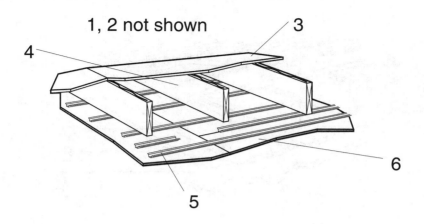

- 1-1/2 in. lightweight concrete or minimum 3/4 in. proprietary gypsum concrete floor topping.

- Building paper (may be optional).

- Nominal 23/32 in. T&G plywood or Oriented strand board (OSB) underlayment (single floor).

- 2x10 wood joists spaced 24 in. o.c.

- Resilient channels.

- 5/8 in. Type X Gypsum Wallboard ceiling.

Fire Tests:

UL R5229-2, 5-25-73, Design L513

Figure M16.1-14 Two-Hour Fire-Resistive Wood Floor/Ceiling Assembly

(2x10 Wood Joists 16" o.c. – Gypsum Directly Applied with Second Layer on Resilient Channels)

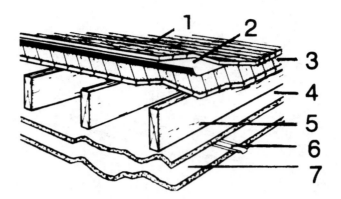

1. Nominal 1 in. wood flooring or 19/32 in. T&G plywood underlayment (single floor). Oriented strand board (OSB) panels are permitted for certain designs.

2. Building paper.

3. Nominal 1x6 T&G boards or 15/32 in. plywood subfloor. Subfloors for certain designs may be nominal 7/16 in. OSB.

4. 5/8 in. proprietary Type X Gypsum Wallboard ceiling attached directly to joists.

5. 2x10 wood joists space 16 in. o.c.

6. Resilient channels.

7. 5/8 in. proprietary Type X Gypsum Wallboard ceiling attached to resilient channels.

Fire Tests:

UL R1319-114, 7-21-67, Design L511

UL R2717-35, 10-21-64, Design L505; ULC Desig
M503

Structural Glued Laminated Timber

Fires do not normally start in structural framing, but rather in the building's contents. These fires generally reach temperatures of between 1,290°F and 1,650°F. Glued laminated timber members perform very well under these conditions. Unprotected steel members typically suffer severe buckling and twisting during fires, often collapsing catastrophically.

Wood ignites at about 480°F, but charring may begin as low as 300°F. Wood typically chars at 1/40 in. per minute. Thus, after 30 minutes of fire exposure, only the outer 3/4 in. of the structural glued laminated timber will be damaged. Char insulates a wood member and hence raises the temperature it can withstand. Most of the cross section will remain intact, and the member will continue supporting loads during a typical building fire.

It is important to note that neither building materials alone, nor building features alone, nor detection and fire extinguishing equipment alone can provide adequate safety from fire in buildings. To ensure a safe structure in the event of fire, authorities base fire and building code requirements on research and testing, as well as fire histories. The model building codes classify Heavy Timber as a specific type of construction and give minimum sizes for roof and floor beams.

The requirements set out for Heavy Timber construction in model building codes do not constitute 1-hour fire resistance. However, procedures are available to calculate the structural glued laminated timber size required for projects in which 1-hour fire resistance is required (see *NDS* 16.2 and AF&PA's *Technical Report 10* available at www.awc.org). The minimum depths for selected structural glued laminated timber sizes that can be adopted for 1-hour fire ratings are given in Table M16.1-6 for structural glued laminated timber beams.

To achieve a 1-hour fire rating for beams whose dimensions qualify them for this rating, the basic layup must be modified – one core lamination must be removed from the center and the tension face augmented with the addition of a tension lamination. For more information concerning the effects of fire on structural glued laminated timber, refer to APA EWS Technical Note Y245 or AITC Technical Note 7. For determining fire resistance other than 1 hour, see *NDS* 16.2 and AF&PA's *Technical Report 10* available at www.awc.org.

Table M16.1-6 Minimum Depths at Which Selected Beam Sizes Can Be Adopted for One-Hour Fire Ratings[1]

Beam Width (in.)	3 Sides Exposed	4 Sides Exposed
6-3/4	13-1/2 or 13-3/4	27 or 27-1/2
8-1/2	7-1/2 or 8-1/4	15 or 15-1/8
8-3/4	6-7/8 or 7-1/2	13-1/2 or 13-3/4
10-1/2	6 or 6-7/8	12 or 12-3/8
10-3/4	6 or 6-7/8	12 or 12-3/8

1. Assuming a load factor of 1.0 (design loads are equal to the capacity of the member). The minimum depths may be reduced when the design loads are less than the member capacity.

M16: FIRE DESIGN

16

Poles and Piles

Very few elements of modern structures can be called "fire proof." Even in buildings where the major structural members are noncombustible, most of the furnishings are flammable. It is for this reason that much of the attention in modern building codes addresses issues related to containing and limiting fires, rather than simply calling materials combustible and noncombustible.

While this topic is fairly complex for other types of products, fire performance is relatively straightforward for poles and piles. Poles are generally used in cross-sectional sizes that qualify as heavy timber construction in the model building codes. On this basis, timber poles compare favorably with other construction materials in their performance under fire conditions. Piles are generally not exposed to fire conditions unless they extend substantially above the groundline.

tructural Composite Lumber

Engineered wood products have fire resistive characteristics very similar to conventional wood frame members. Since many engineered wood products are proprietary, they are usually recognized in a code evaluation report published by an evaluation service. Each evaluation report usually contains fire resistance information.

Very few elements of modern structures can be called "fire proof." Even in buildings where the major structural members are noncombustible, most of the furnishings are flammable. It is for this reason that much of the attention in model building codes addresses issues related to containing and limiting fires, rather than simply calling materials combustible and noncombustible. The primary intent of the building codes is to ensure structural stability to allow exiting, evacuation, and fire fighting.

As with the previous topic of durability, this Manual cannot cover the topic of designing for optimal structural performance in fire conditions in detail. There are many excellent texts on the topic, and designers are advised to use this information early in the design process. To assist the designer in understanding several ways in which fire performance can be addressed, the following overview is provided.

Fire sprinklers are probably the most effective method to enhance fire resistance of engineered wood systems (as well as other systems). They are designed to control the fire while protecting the occupants and the building until the fire department arrives. They are the ultimate way to improve fire safety.

Heavy timber construction has proven to be acceptable in many areas where fire safety is of utmost concern. These applications have proven to be not only reliable, but economical in certain structures – many wider width SCL products can be used in heavy timber construction. Consult manufacturer's literature or code evaluation reports for specific information.

The fire performance of wood structures can be enhanced in the same ways as that of structures of steel, concrete, or masonry:

- Fire sprinkler systems have proven to be effective in a variety of structures, both large and small
- Protection of the structural members with materials such as properly attached gypsum sheathing can provide greatly improved fire performance. Fire ratings, as established from test procedures specified in ASTM E-119, of up to 2 hours can be achieved through the use of gypsum sheathing
- Where surface burning characteristics are critical, fire-retardant treatments can be used to reduce the flamespread for some products

To reiterate, this Manual does not purport to address this topic in an all-inclusive manner – it is merely an attempt to alert designers to the need to address fire performance issues in the design of the structure.

Wood I-Joists

The wood I-joist industry has actively supported the following projects to establish fire performance of systems using wood I-joist products:

- ASTM E-119 fire tests have been conducted by the wood I-joist industry to establish fire resistance ratings for generic I-joist systems. Detailed descriptions of these systems are shown in Figures M16.1-15 through M16.1-21 and are summarized in Table M16.1-7. In addition, sound transmission class (STC) and impact insulation class (IIC) ratings for each of these assemblies are provided. Updates to this information will be posted on the American Wood Council's website at www.awc.org.

- ASTM E-119 fire tests have been conducted by wood I-joist manufacturers to establish fire resistance ratings for proprietary I-joist systems. Detailed descriptions of these systems are available from the individual I-joist manufacturer.

- National Fire Protection Research Foundation Report titled "National Engineered Lightweight Construction Fire Research Project." This report documents an extensive literature search of the fire performance of engineered lightweight construction.

- A video, *I-JOISTS: FACTS ABOUT PROGRESS*, has been produced by the Wood I-Joist Manufacturer's Association (WIJMA). This video describes some basic facts about changes taking place within the construction industry and the fire service. Along with this video is a document that provides greater details on fire performance issues.

- Industry research in fire endurance modeling for I-joist systems.

Fire-Resistive Wood I-Joist Floor/Ceiling Assemblies

One-Hour Assemblies

Joists	Insulation	Furring		Ceiling Sheathing	Fasteners	Details
I-joists @ 24" o.c. Min. flange thickness: 1-1/2" Min. flange area: 5.25 sq. in. Min. web thickness: 3/8" Min. I-joist depth: 9-1/4"	1-1/2" mineral wool batts (2.5 pcf - nominal) Resting on hat-shaped channels	Hat-shaped channels	F	5/8" Type C Gypsum Wallboard	1-1/8" Type S drywall screws @ 12" o.c. *(see fastening details)*	Figure M16.1-15
I-joists @ 24" o.c. Min. flange thickness: 1-1/2" Min. flange area: 5.25 sq. in. Min. web thickness: 7/16" Min. I-joist depth: 9-1/4"	1-1/2" mineral wool batts (2.5 pcf - nominal) Resting on resilient channels	Resilient channels	F	5/8" Type C Gypsum Wallboard	1" Type S drywall screws @ 8" o.c. *(see fastening details)*	Figure M16.1-16
I-joists @ 24" o.c. Min. flange thickness: 1-5/16" Min. flange area: 2.25 sq. in. Min. web thickness: 3/8" Min. I-joist depth: 9-1/4"	2" mineral wool batts (3.5 pcf - nominal) Resting on 1x4 setting strips	Resilient channels	F	5/8" Type C Gypsum Wallboard	1-1/8" Type S drywall screws @ 7" o.c. *(see fastening details)*	Figure M16.1-17
I-joists @ 24" o.c. Min. flange thickness: 1-1/2" Min. flange area: 3.45 sq. in. Min. web thickness: 3/8" Min. I-joist depth: 9-1/4"	1" mineral wool batts (6 pcf - nominal) Resting on hat-shaped channels under I-joist bottom flange	Hat-shaped channels supported by CSC clips	F	1/2" Type C Gypsum Wallboard	1" Type S drywall screws @ 12" o.c. *(see fastening details)*	Figure M16.1-18
I-joists @ 24" o.c. Min. flange thickness: 1-1/2" Min. flange area: 2.25 sq. in. Min. web thickness: 3/8" Min. I-joist depth: 9-1/4"	(none)	(none)	B	1/2" Type X Gypsum Wallboard	1-5/8" Type S drywall screws @ 12" o.c.	Figure M16.1-19
			F	1/2" Type X Gypsum Wallboard	2" Type S drywall screws @ 12" o.c. 1-1/2" Type G drywall screws @ 8" o.c. (see fastening details)	
I-joists @ 24" o.c. Min. flange thickness: 1-5/16" Min. flange area: 1.95 sq. in. Min. web thickness: 3/8" Min. I-joist depth: 9-1/2"	(optional)	Resilient channels	B	1/2" Type X Gypsum Wallboard	1-1/4" Type S drywall screws @ 12" o.c.	Figure M16.1-20
			F	1/2" Type X Gypsum Wallboard	1-5/8" Type S drywall screws @ 12" o.c. 1-1/2" Type G drywall screws @ 8" o.c. (see fastening details)	

Two-Hour Assembly

Joists	Insulation	Furring		Ceiling Sheathing	Fasteners	Details
I-joists @ 24" o.c. Min. flange thickness: 1-1/2" Min. flange area: 2.25 sq. in. Min. web thickness: 3/8" Min. I-joist depth: 9-1/4"	3-1/2" fiberglass insulation supported by stay wires spaced 12" o.c.	(none)	B	5/8" Type C Gypsum Wallboard	1-5/8" Type S drywall screws @ 12" o.c.	Figure M16.1-21
		Hat-shaped channels	M	5/8" Type C Gypsum Wallboard	1" Type S drywall screws @ 12" o.c.	
			F	5/8" Type C Gypsum Wallboard	1-5/8" Type S drywall screws @ 8" o.c. *(see fastening details)*	

B- Base layer sheathing (direct attached)   **M**- Middle layer sheathing   **F**- Face layer sheathing

Figure M16.1-15   One-Hour Fire-Resistive Ceiling Assembly (WIJ-1.1)
Floor[a]/Ceiling - 100% Design Load - 1-Hour Rating - ASTM E 119/NFPA 251

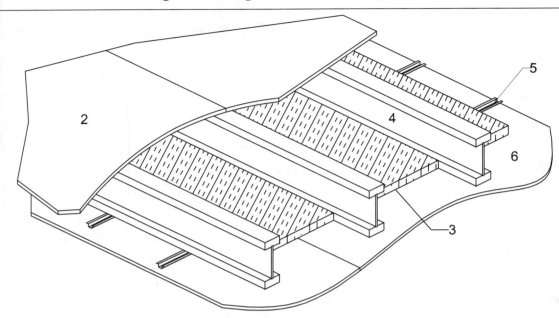

1. **Floor Topping (optional, not shown):** Gypsum concrete, lightweight or normal concrete topping.
2. **Floor Sheathing:** Minimum 23/32-inch-thick tongue-and-groove wood sheathing (Exposure 1). Installed per code requirements with minimum 8d common nails and glued to joist top flanges with AFG-01 construction adhesive.
3. **Insulation:** Minimum 1-1/2-inch-thick mineral fiber insulation batts – 2.5 pcf (nominal), supported by furring channels.
4. **Structural Members:** Wood I-joists spaced a maximum of 24 inches on center.
    Minimum I-joist flange depth: 1-1/2 inches   Minimum I-joist flange area: 5.25 inches[2]
    Minimum I-joist web thickness: 3/8 inch   Minimum I-joist depth: 9-1/4 inches

Types of Adhesives Used in Tested I-Joists		
Flange-to-Flange Endjoint	**Flange-to-Web Joint**	**Web-to-Web Endjoint**
Resorcinol-Formaldehyde	Phenol-Resorcinol-Formaldehyde	Phenol-Resorcinol-Formaldehyde

5. **Furring Channels:** Minimum 0.026-inch-thick galvanized steel hat-shaped furring channels, attached perpendicular to I-joists using 1-5/8-inch-long drywall screws. Furring channels spaced 16 inches on center and doubled at each wallboard end joint extending to the next joist.
6. **Gypsum Wallboard:** Minimum 5/8-inch-thick Type C gypsum wallboard installed with long dimension perpendicular to furring channels and fastened to each channel with minimum 1-1/8-inch-long Type S drywall screws. Fasteners spaced 12 inches on center in the field of the wallboard, 8 inches on center at wallboard end joints, and 3/4 inch from panel edges and ends. End joints of wallboard staggered.
7. **Finish System (not shown):** Face layer joints covered with tape and coated with joint compound. Screw heads covered with joint compound.

Fire Test conducted at Gold Bond Building Products Research Center: February 9, 1990
Third-Party Witness: Warnock Hersey International, Inc.   Report No: WHI-651-0311.1

STC and IIC Sound Ratings for Listed Assembly							
Without Gypsum Concrete				**With Gypsum Concrete**			
Cushioned Vinyl		**Carpet & Pad**		**Cushioned Vinyl**		**Carpet & Pad**	
STC	**IIC**	**STC**	**IIC**	**STC**	**IIC**	**STC**	**IIC**
-	-	-	-	-	-	49[b]	59[b]

a. This assembly may also be used in a fire-resistive roof/ceiling application, but only when constructed exactly as described.
b. STC and IIC values estimated by David L. Adams Associates, Inc.

Figure M16.1-16 One-Hour Fire-Resistive Ceiling Assembly (WIJ-1.2)

Floor[a]/Ceiling - 100% Design Load - 1 Hour Rating - ASTM E 119/NFPA 251

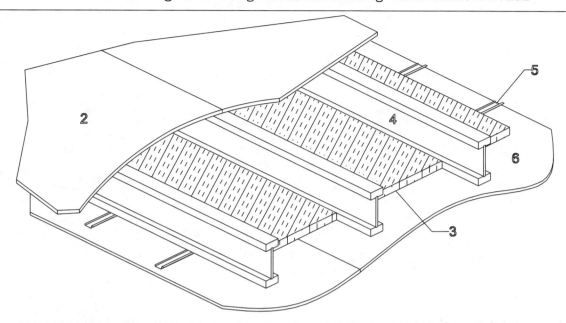

1. **Floor Topping (optional, not shown):** Gypsum concrete, lightweight or normal concrete topping.
2. **Floor Sheathing:** Minimum 23/32-inch-thick tongue-and-groove wood sheathing (Exposure 1). Installed per code requirements with minimum 8d common nails, and glued to joist top flanges with AFG-01 construction adhesive.
3. **Insulation:** Minimum 1-1/2-inch-thick mineral fiber insulation batts – 2.5 pcf (nominal), supported by resilient channels.
4. **Structural Members:** Wood I-joists spaced a maximum of 24 inches on center.

Minimum I-joist flange depth: 1-1/2 inches	Minimum I-joist flange area: 5.25 inches²
Minimum I-joist web thickness: 7/16 inch	Minimum I-joist depth: 9-1/4 inches

Types of Adhesives Used in Tested I-Joists		
Flange-to-Flange Endjoint	**Flange-to-Web Joint**	**Web-to-Web Endjoint**
Phenol-Resorcinol-Formaldehyde	Phenol-Resorcinol-Formaldehyde	Phenol-Resorcinol-Formaldehyde

5. **Resilient Channels:** Minimum 0.019-inch-thick galvanized steel resilient channels, attached perpendicular to I-joists using 1-5/8-inch-long drywall screws. Resilient channels spaced 16 inches on center and doubled at each wallboard end joint extending to the next joist.
6. **Gypsum Wallboard:** Minimum 5/8-inch-thick Type C gypsum wallboard installed with long dimension perpendicular to resilient channels and fastened to each channel with minimum 1-inch-long Type S drywall screws. Fasteners spaced 12 inches on center in the field of the wallboard, 8 inches on center at wallboard end joints, and 3/4 inch from panel edges and ends. End joints of wallboard staggered.
7. **Finish System (not shown):** Face layer joints covered with tape and coated with joint compound. Screw heads covered with joint compound.

Fire Test conducted at Gold Bond Building Research Center:	June 19, 1984
Third-Party Witness: Warnock Hersey International, Inc.	Report No: WHI-694-0159

STC and IIC Sound Ratings for Listed Assembly							
Without Gypsum Concrete				**With Gypsum Concrete**			
Cushioned Vinyl		**Carpet & Pad**		**Cushioned Vinyl**		**Carpet & Pad**	
STC	**IIC**	**STC**	**IIC**	**STC**	**IIC**	**STC**	**IIC**
51[b]	46[b]	51[b]	64[b]	60[b]	50[b]	60[b]	65[b]

a. This assembly may also be used in a fire-resistive roof/ceiling application, but only when constructed exactly as described.
b. STC and IIC values estimated by David L. Adams Associates, Inc.

M16: FIRE DESIGN

16

Figure M16.1-17 One-Hour Fire-Resistive Ceiling Assembly (WIJ-1.3)

Floor[a]/Ceiling - 100% Design Load - 1-Hour Rating - ASTM E 119/NFPA 251

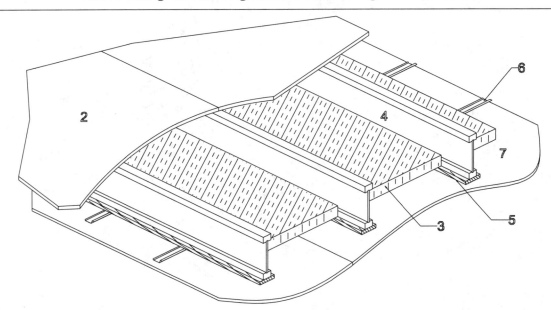

1. **Floor Topping (optional, not shown):** Gypsum concrete, lightweight or normal concrete topping.
2. **Floor Sheathing:** Minimum 23/32-inch-thick tongue-and-groove wood sheathing (Exposure 1). Installed per code requirements.
3. **Insulation:** Minimum 2-inch-thick mineral fiber insulation batts – 3.5 pcf (nominal), supported by setting strips edges friction-fitted between the sides of the I-joist flanges.
4. **Structural Members:** Wood I-joists spaced a maximum of 24 inches on center.

   Minimum I-joist flange depth: 1-5/16 inches   Minimum I-joist flange area: 2.25 inches^2

   Minimum I-joist web thickness: 3/8 inch   Minimum I-joist depth: 9-1/4 inches

Types of Adhesives Used in Tested I-Joists		
LVL Flange Adhesive	**Flange-to-Web Joint**	**Web-to-Web Endjoint**
Phenol-Resorcinol-Formaldehyde	Emulsion Polymer Isocyanate	Polyurethane Emulsion Polymer

5. **Setting Strips:** Nominal 1x4 wood setting strips attached with 1-1/2-inch-long drywall screws at 24 inches on center along the bottom flange of I-joist creating a ledge to support insulation.
6. **Resilient Channels:** Minimum 0.019-inch-thick galvanized steel resilient channels, attached perpendicular to I-joists using 1-7/8-inch-long drywall screws. Resilient channels spaced 16 inches on center and doubled at each wallboard end joint extending to the next joist.
7. **Gypsum Wallboard:** Minimum 5/8-inch-thick Type C gypsum wallboard installed with long dimension perpendicular to resilient channels and fastened to each channel with minimum 1-1/8-inch-long Type S drywall screws. Fasteners spaced 7 inches on center and 3/4 inch from panel edges and ends. End joints of wallboard staggered.
8. **Finish System (not shown):** Face layer joints covered with tape and coated with joint compound. Screw heads covered with joint compound.

   Fire Test conducted at National Gypsum Testing Services, Inc.:   September 28, 2001

   Third-Party Witness: Underwriter's Laboratories, Inc.   Report No: NC3369

STC and IIC Sound Ratings for Listed Assembly							
Without Gypsum Concrete				With Gypsum Concrete			
Cushioned Vinyl		Carpet & Pad		Cushioned Vinyl		Carpet & Pad	
STC	IIC	STC	IIC	STC	IIC	STC	IIC
51[b]	46[b]	52	66	60[b]	48[b]	60[b]	60[b]

a. This assembly may also be used in a fire-resistive roof/ceiling application, but only when constructed exactly as described.
b. STC and IIC values estimated by David L. Adams Associates, Inc.

Figure M16.1-18 One-Hour Fire-Resistive Ceiling Assembly (WIJ-1.4)

Floor[a]/Ceiling - 100% Design Load - 1-Hour Rating - ASTM E 119/NFPA 251

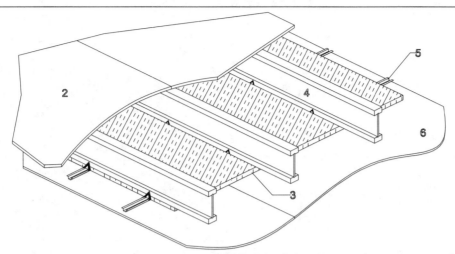

. **Floor Topping (optional, not shown):** Gypsum concrete, lightweight or normal concrete topping.

. **Floor Sheathing:** Minimum 23/32-inch-thick tongue-and-groove wood sheathing (Exposure 1). Installed per code requirements with minimum 8d common nails.

. **Insulation:** Minimum 1-inch-thick mineral fiber insulation batts – 6 pcf (nominal) with width equal to the on-center spacing of the I-joists. Batts installed on top of furring channels and under bottom flange of I-joists with the sides butted against support clips. Abutted ends of batts centered over furring channels with batts tightly butted at all joints.

. **Structural Members:** Wood I-joists spaced a maximum of 24 inches on center.

Minimum I-joist flange depth: 1-1/2 inches	Minimum I-joist flange area: 3.45 inches^2
Minimum I-joist web thickness: 3/8 inch	Minimum I-joist depth: 9-1/4 inches

Types of Adhesives Used in Tested I-Joists		
Flange-to-Flange Endjoint	**Flange-to-Web Joint**	**Web-to-Web Endjoint**
Phenol-Resorcinol-Formaldehyde	Phenol-Resorcinol-Formaldehyde	Resorcinol-Formaldehyde

. **Furring Channels:** Minimum 0.019-inch-thick galvanized steel hat-shaped furring channels, attached perpendicular to I-joists spaced 24 inches on center. At channel splices, adjacent pieces overlap a minimum of 6 inches and tied with a double strand of No. 18 gage galvanized steel wire at each end of the overlap. Channels secured to I-joists with Simpson Type CSC support clips at each intersection with the I-joists. Clips nailed to the side of I-joist bottom flange with one 1-1/2-inch-long No. 11 gage nail. A row of furring channel located on each side of wallboard end joints and spaced 2.25 inches from the end joint (4.5 inches on center).

. **Gypsum Wallboard:** Minimum 1/2-inch-thick Type C gypsum wallboard. Wallboard installed with long dimension perpendicular to furring channels and fastened to each channel with minimum 1-inch-long Type S drywall screws. Fasteners spaced 12 inches on center in the field of the wallboard, 6 inches on center at the end joints, and 3/4 inch from panel edges and ends. End joints of wallboard continuous or staggered. For staggered wallboard end joints, furring channels extend a minimum of 6 inches beyond each end of the joint.

. **Finish System (not shown):** Face layer joints covered with tape and coated with joint compound. Screw heads covered with joint compound.

Fire Test conducted at Underwriter's Laboratories, Inc.　　May 11, 1983
Third-Party Witness: Underwriter's Laboratories, Inc.　　Report No: UL R1037-1

STC and IIC Sound Ratings for Listed Assembly							
Without Gypsum Concrete				**With Gypsum Concrete**			
Cushioned Vinyl		**Carpet & Pad**		**Cushioned Vinyl**		**Carpet & Pad**	
STC	**IIC**	**STC**	**IIC**	**STC**	**IIC**	**STC**	**IIC**
-	-	46	68	51	47	50	73

. This assembly may also be used in a fire-resistive roof/ceiling application, but only when constructed exactly as described.

Figure M16.1-19 One-Hour Fire-Resistive Ceiling Assembly (WIJ-1.5)
Floor[a]/Ceiling - 100% Design Load - 1-Hour Rating - ASTM E 119/NFPA 251

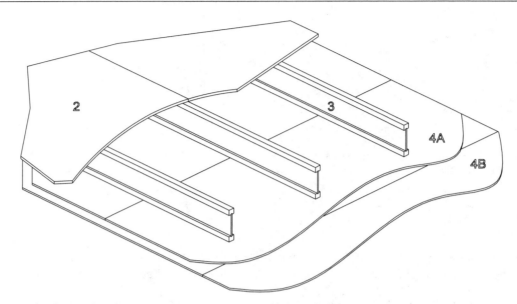

1. **Floor Topping (optional, not shown):** Gypsum concrete, lightweight or normal concrete topping.
2. **Floor Sheathing:** Minimum 23/32-inch-thick tongue-and-groove wood sheathing (Exposure 1). Installed per code requirements with minimum 8d common nails.
3. **Structural Members:** Wood I-joists spaced a maximum of 24 inches on center.

Minimum I-joist flange depth: 1-1/2 inches	Minimum I-joist flange area: 2.25 inches^2
Minimum I-joist web thickness: 3/8 inch	Minimum I-joist depth: 9-1/4 inches

Types of Adhesives Used in Tested I-Joists		
LVL Flange Adhesive	**Flange-to-Web Joint**	**Web-to-Web Endjoint**
Phenol-Resorcinol-Formaldehyde	Phenol-Resorcinol-Formaldehyde	Phenol-Resorcinol-Formaldehyde

4. **Gypsum Wallboard:** Two layers of minimum 1/2- inch Type X gypsum wallboard attached with the long dimension perpendicular to the I-joists as follows:

   4a. **Wallboard Base Layer:** Base layer of wallboard attached to bottom flange of I-joists using 1-5/8-inch Type drywall screws at 12 inches on center. End joints of wallboard centered on bottom flange of the I-joist and staggered.

   4b. **Wallboard Face Layer:** Face layer of wallboard attached to bottom flange of I-joists through base layer using 2 inch Type S drywall screws spaced 12 inches on center on intermediate joists and 8 inches on center at end joints. Edg joints of wallboard face layer offset 24 inches from those of base layer. End joints centered on bottom flange of I-joist and offset a minimum of one joist spacing from those of base layer. Additionally, face layer of wallboard attached to bas layer with 1-1/2-inch Type G drywall screws spaced 8 inches on center, placed 6 inches from face layer end joints.

5. **Finish System (not shown):** Face layer joints covered with tape and coated with joint compound. Screw heads covered with joint compound.

Fire Tests conducted at Factory Mutual Research: September 29, 1978
Third-Party Witness: Factory Mutual Research: Report No: FC-268
PFS Test Report #86-09-1:   July 28, 1986

STC and IIC Sound Ratings for Listed Assembly							
Without Gypsum Concrete				**With Gypsum Concrete**			
Cushioned Vinyl		**Carpet & Pad**		**Cushioned Vinyl**		**Carpet & Pad**	
STC	**IIC**	**STC**	**IIC**	**STC**	**IIC**	**STC**	**IIC**
-	-	-	-	-	-	49 [b]	55 [b]

a. This assembly may also be used in a fire-resistive roof/ceiling application, but only when constructed exactly as described.
b. STC and IIC values estimated by David L. Adams Associates, Inc.

Figure M16.1-20 One-Hour Fire-Resistive Ceiling Assembly (WIJ-1.6)

Floor[a]/Ceiling - 100% Design Load - 1-Hour Rating - ASTM E 119/NFPA 251

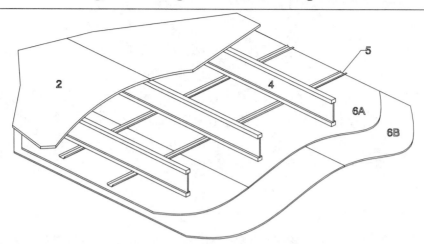

. **Floor Topping (optional, not shown):** Gypsum concrete, lightweight or normal concrete topping.

. **Floor Sheathing:** Minimum 23/32-inch-thick tongue-and-groove wood sheathing (Exposure 1). Installed per code requirements with minimum 8d common nails.

. **Insulation (optional, not shown):** Insulation fitted between I-joists supported by the resilient channels.

. **Structural Members:** Wood I-joists spaced a maximum of 24 inches on center.

    Minimum I-joist flange depth: 1-5/16 inches   Minimum I-joist flange area: 1.95 inches^2

    Minimum I-joist web thickness: 3/8 inch   Minimum I-joist depth: 9-1/2 inches

Types of Adhesives Used in Tested I-Joists		
LVL Flange Adhesive	**Flange-to-Web Joint**	**Web-to-Web Endjoint**
Phenol-Resorcinol-Formaldehyde	Emulsion Polymer Isocyanate	Emulsion Polymer Isocyanate

. **Resilient Channels[b]:** Minimum 0.019-inch-thick galvanized steel resilient channel attached perpendicular to the bottom flange of the I-joists with one 1-1/4-inch drywall screw. Channels spaced a maximum of 16 inches on center [24 inches on center when I-joists are spaced a maximum of 16 inches on center].

. **Gypsum Wallboard:** Two layers of minimum 1/2-inch Type X gypsum wallboard attached with the long dimension perpendicular to the resilient channels as follows:

6a. **Wallboard Base Layer:** Base layer of wallboard attached to resilient channels using 1-1/4-inch Type S drywall screws at 12 inches on center.

6b. **Wallboard Face Layer:** Face layer of wallboard attached to resilient channels through base layer using 1-5/8-inch Type S drywall screws spaced 12 inches on center. Edge joints of wallboard face layer offset 24 inches from those of base layer. Additionally, face layer of wallboard attached to base layer with 1-1/2-inch Type G drywall screws spaced 8 inches on center, placed 6 inches from face layer end joints.

. **Finish System (not shown):** Face layer joints covered with tape and coated with joint compound. Screw heads covered with joint compound.

Fire Test conducted at National Research Council of Canada: Report No: A-4440.1 June 24, 1997

	STC and IIC Sound Ratings for Listed Assembly							
	Without Gypsum Concrete				**With Gypsum Concrete**			
	Cushioned Vinyl		**Carpet & Pad**		**Cushioned Vinyl**		**Carpet & Pad**	
	STC	**IIC**	**STC**	**IIC**	**STC**	**IIC**	**STC**	**IIC**
With Insulation	59	50	55 [c]	68 [c]	65	51	63 [c]	65 [c]
Without Insulation	-	-	54	68	-	-	58 [c]	55 [c]

. This assembly may also be used in a fire-resistive roof/ceiling application, but only when constructed exactly as described.

. Direct attachment of gypsum wallboard in lieu of attachment to resilient channels is typically deemed acceptable. When gypsum wallboard is directly attached to the I-joists, the wallboard should be installed with long dimension perpendicular to the I-joists. When insulation is used, it should be fitted between I-joists, supported above the wallboard at the spacing specified for resilient channels.

. STC and IIC values estimated by David L. Adams Associates, Inc.

Figure M16.1-21 Two-Hour Fire-Resistive Ceiling Assembly (WIJ-2.1)

Floor[a]/Ceiling - 100% Design Load - 2-Hour Rating - ASTM E119/NFPA 251

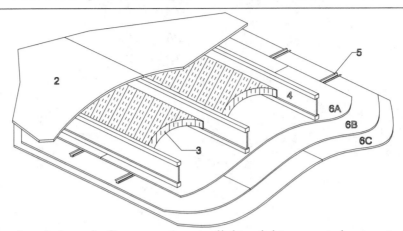

1. **Floor Topping (optional, not shown):** Gypsum concrete, lightweight or normal concrete topping.
2. **Floor Sheathing:** Minimum 23/32-inch-thick tongue-and-groove wood sheathing (Exposure 1). Installed per code requirements.
3. **Insulation:** Minimum 3-1/2-inch-thick unfaced fiberglass insulation fitted between I-joists supported by stay wire spaced 12 inches on center.
4. **Structural Members:** Wood I-joists spaced a maximum of 24 inches on center.

Minimum I-joist flange depth: 1-1/2 inches   Minimum I-joist flange area: 2.25 inches^2

Minimum I-joist web thickness: 3/8 inch   Minimum I-joist depth: 9-1/4 inches

Types of Adhesives Used in Tested I-Joists		
LVL Flange Adhesive	**Flange-to-Web Joint**	**Web-to-Web Endjoint**
Phenol-Resorcinol-Formaldehyde	Phenol-Resorcinol-Formaldehyde	Phenol-Resorcinol-Formaldehyde

5. **Furring Channels:** Minimum 0.0179-inch-thick galvanized steel hat-shaped furring channels, attached perpendicular to I-joists using 1-5/8-inch long drywall screws. Furring channels spaced 16 inches on center (furring channels used to support the second and third layers of gypsum wallboard).
6. **Gypsum Wallboard:** Three layers of minimum 5/8-inch Type C gypsum wallboard as follows:
   6a. **Wallboard Base Layer:** Base layer of wallboard attached to bottom flange of I-joists using 1-5/8-inch Type S drywall screws at 12 inches on center with the long dimension of wallboard perpendicular to I-joist. End joints of wallboard centered on bottom flange of the I-joist and staggered from end joints in adjacent sheets.
   6b. **Wallboard Middle Layer:** Middle layer of wallboard attached to furring channels using 1-inch Type S drywall screws spaced 12 inches on center with the long dimension of wallboard perpendicular to furring channels. End joints staggered from end joints in adjacent sheets.
   6c. **Wallboard Face Layer:** Face layer of wallboard attached to furring channels through middle layer using 1-5/8-inch Type S drywall screws spaced 8 inches on center. Edge joints of wallboard face layer offset 24 inches from those of middle layer. End joints of face layer of wallboard staggered with respect to the middle layer.
7. **Finish System (not shown):** Face layer joints covered with tape and coated with joint compound. Screw heads covered with joint compound.

Fire Test conducted at Gold Bond Building Products Research Center: December 16, 1992

Third-Party Witness: PFS Corporation: Report No: #92-56

STC and IIC Sound Ratings for Listed Assembly							
Without Gypsum Concrete				**With Gypsum Concrete**			
Cushioned Vinyl		**Carpet & Pad**		**Cushioned Vinyl**		**Carpet & Pad**	
STC	**IIC**	**STC**	**IIC**	**STC**	**IIC**	**STC**	**IIC**
-	-	49[b]	54[b]	52[b]	46[b]	52[b]	60[b]

a. This assembly may also be used in a fire-resistive roof/ceiling application, but only when constructed exactly as described.
b. STC and IIC values estimated by David L. Adams Associates, Inc.

Metal Plate Connected Wood Trusses

Generally, a fire endurance rating of 1-hour is mandated by code for many of the applications where trusses would be used. All testing on these assemblies is performed in accordance with the ASTM's *Standard Methods for Fire Tests of Building Construction and Materials* (ASTM C119).

The two primary source documents for fire endurance assembly results are the *Fire Resistance Design Manual*, published by the Gypsum Association (GA) and the *Fire Resistance Directory,* published by Underwriters' Laboratories, Inc. (UL). Warnock Hersey (WH) assemblies are now listed in the *ITS Directory of Listed Products*. These tested assemblies are available for specification by Architects or Building Designers, and for use by all Truss Manufacturers where a rated assembly is required, and can generally be applied to both floor and roof assembly applications.

According to the UL Directory's Design Information section: "Ratings shown on individual designs apply to equal or greater height or thickness of the assembly, and to larger structural members, when both size and weight are equal or larger than specified, and when the thickness of the flanges, web or diameter of chords is equal or greater." Thus, larger and deeper trusses can be used under the auspices of the same design number. This approach has often been used in roof truss applications since roof trusses are usually much deeper than the tested assemblies.

Thermal and/or acoustical considerations at times may require the installation of insulation in a floor-ceiling or roof-ceiling assembly that has been tested without insulation. As a general 'rule,' experience indicates that it is allowable to add insulation to an assembly, provided that the depth of the truss is increased by the depth of the insulation. And as a general 'rule,' assemblies that were tested with insulation may have the insulation removed.

To make a rational assessment of any modification to a tested assembly, one must look at the properties of the insulation and the impact that its placement inside the assembly will have on the fire endurance performance of the assembly. Insulation retards the transfer of heat, is used to retain heat in warm places, and reduces the flow of heat into colder areas. As a result, its addition to a fire endurance assembly will affect the flow of heat through and within an assembly. One potential effect of insulation placed directly on the gypsum board is to retard the dissipation of heat through the assembly, concentrating heat in the protective gypsum board.

In some cases specific branded products are listed in the test specifications. Modifications or substitutions to fire endurance assemblies should be reviewed with the building designer and code official, preferably with the assistance of a professional engineer. This review is required because the final performance of the assembly is a result of the composite of the materials used in the construction of the assembly.

The following pages, courtesy of WTCA – Representing the Structural Building Components Industry, include brief summaries of wood truss fire endurance assemblies and sound transmission ratings. For more information, visit www.sbcindustry.com. Also, several truss plate manufacturers have developed proprietary fire resistant assemblies. These results apply only to the specific manufacturer's truss plates and referenced fire endurance assembly system. For more detailed information on these assemblies, the individual truss plate manufacturer should be contacted. Complete specifications on the UL, GA, and WH assemblies are available at their respective websites.

Roof and Floor Assemblies

The following are only summaries. Users must consult the listed testing agency's documentation for complete information.

Certifying Agencies:

GA = Gypsum Association
NER = National Evaluation Service Report
PFS = PFS Corporation
UL = Underwriters Laboratory
WH = Warnock Hersey International, Ltd.

Note:

In some cases specific branded products are listed in the test specifications. There are situations where comparable products <u>may</u> be substituted.

⊠	Wood blocking at gypsum joints
⎿⏌	FR-Quik Channel Sets at gypsum joints
⌣	Furring channel
⌇	Resilient channel

45-Minute Fire Resistive Truss Designs:

PFS 88-03, FR-SYSTEM 4
(floor or roof - optional insulation)

Fire Rating: 45 Minutes
Finish Rating: 22 Minutes
Construction: Wood trusses 24" oc, flat or floor minimum
depth 15"
Sloped minimum pitch 3/12, depth 19-1/2"
FR-Quik Channel Sets™ and Bond
Washers™ by Alpine Engineered Products
One layer 5/8" Type C gypsum board
Sheathing minimum 15/32"

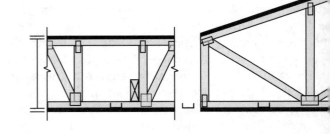

WH TSC/FCA 45-02
(floor or roof - optional insulation)

Fire Rating: 45 Minutes
Finish Rating: 22 Minutes
Construction: Wood trusses 24" oc, minimum 10" depth
Truswal metal truss plates
One layer 5/8" Type X gypsum board
Sheathing minimum 3/4"

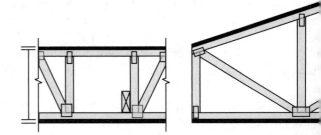

WH TSC/FCA 45-04
(floor or roof - suspended ceiling)

Fire Rating: 45 Minutes
Finish Rating: 22 Minutes
Construction: Wood trusses 24" oc, minimum 10" depth
Truswal metal truss plates
Fire rated suspended ceiling - a minimum of
7-1/2" below the joist
Sheathing minimum 3/4"

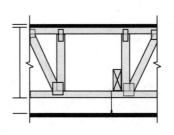
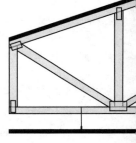

One-Hour Fire Resistive Truss Designs:

GA – FC5406 & FC5408 RC2601 & RC2602
(floor or roof) (see also *IBC* Table 720.1(3)

Fire Rating: 1 Hr
Finish Rating: unknown
Construction: Wood trusses 24" oc, minimum 9-1/4" depth
Two layers 5/8" Type X gypsum board
Sheathing minimum 1/2"

GA – FC5512
(floor or roof)

Fire Rating: 1 Hr
Finish Rating: unknown
Construction: Wood trusses 24" oc, minimum 12" depth
Two layers 1/2" Type X gypsum board
Sheathing minimum 19/32"

GA – FC5515 or FC5516
(floor or roof)

Fire Rating: 1 Hr
Finish Rating: unknown
Construction: Wood trusses 24" oc, minimum 12" depth
Rigid furring channel 24" oc
One layer 5/8" proprietary type X gypsum board
Sheathing nominal 3/4"

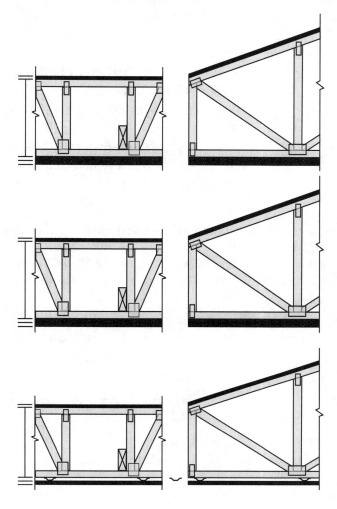

GA – FC5517, PFS 86-10, or TPI/WTCA FC-392
(floor or roof)

Fire Rating: 1 Hr
Finish Rating: unknown
Construction: Wood trusses 24" oc, minimum 14-1/4" depth
Wood blocking secured with metal clips
One layer 5/8" proprietary type X gypsum board
Sheathing nominal 5/8" or 23/32"

ER – 392 WTCA – FR-SYSTEM 1™
(floor or roof - optional insulation)

Fire Rating: 1 Hr
Finish Rating: 23 Minutes
Construction: Wood trusses (nominal 2x3) 24" oc, minimum
16" depth
FR-Quik Channel Sets™ and Bond
Washers™ by Alpine Engineered Products
One layer 5/8" proprietary type X gypsum board
Sheathing minimum 23/32"

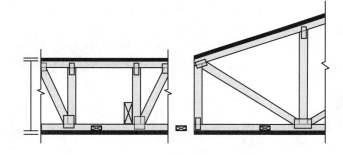

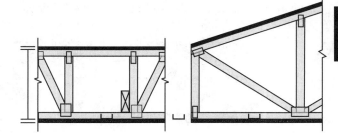

NER – 392 WTCA – FR-SYSTEM 3™
(floor or roof - optional insulation)

Fire Rating: 1 Hr

Finish Rating: unknown

Construction: Wood trusses (nominal 2x3) 24" oc, flat or
floor minimum 15" depth

Sloped minimum pitch 3/12, depth 19-1/2"

FR-Quik Channel Sets™ and Bond
Washers™ by Alpine Engineered Products

Two layers 1/2" type X gypsum board

Sheathing minimum 15/32"

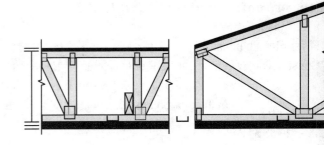

NER – 399
(floor or roof - insulation, suspended ceiling, and light fixtures)
(see *WTCA Metal Plate Connected Wood Truss Handbook* for details – report discontinued)

Fire Rating: 1 Hr

Finish Rating: 35 Minutes

Construction: Wood trusses max. 8' oc, minimum 16" depth

Fire rated suspended ceiling system

Purlins spaced 24" oc

Sheathing minimum 23/32"

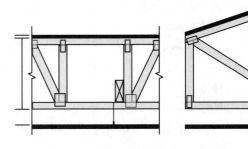

PFS 89-58, FR-SYSTEM 5™
(floor or flat roof - insulation)

Fire Rating: 1 Hr

Finish Rating: 26 Minutes

Construction: Wood trusses 24" oc, minimum depth 10"

2" nominal shield member

FR-Quik Channel Sets™ and Bond
Washers™ by Alpine Engineered Products

One layer 5/8" proprietary Type X gypsum
board

Sheathing minimum 23/32"

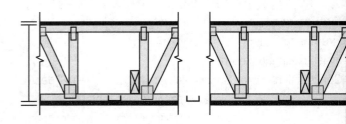

UL – L528 & L534
(floor or roof)

Fire Rating: 1 Hr

Finish Rating: 22 Minutes

Construction: Wood trusses 24" oc, minimum 12" depth for
L528, 18" for L534

Furring channel 24" oc, alt. resilient channel
16" oc

One layer 5/8" proprietary Type X gypsum
board

Sheathing minimum 23/32"

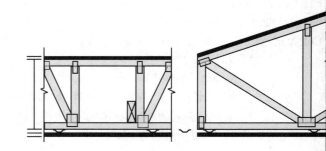

UL – L529
(floor or roof - dropped ceiling, damper, duct, fixtures & metal trim)

Fire Rating: 1 Hr

Finish Rating: 22 Minutes

Construction: Wood trusses 24" oc, minimum 12" depth

Ceiling system dropped 7-1/2"

One layer 5/8" proprietary Type X gypsum board

Sheathing minimum 23/32"

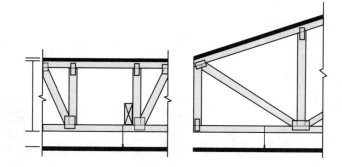

UL – L542
(floor or roof)

Fire Rating: 1 Hr

Finish Rating: unknown

Construction: Wood trusses 24" oc, minimum 12" depth

Two layers 1/2" proprietary Type X gypsum board

Sheathing minimum 23/32"

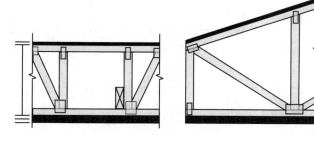

UL – L546
(floor or roof – air duct & damper, optional insulation)

Fire Rating: 1 Hr

Finish Rating: 25 Minutes

Construction: Wood trusses 24" oc, minimum 18" depth

Resilient channel 16" oc or 12" oc

One layer 5/8" proprietary Type X gypsum board

Sheathing minimum 15/32"

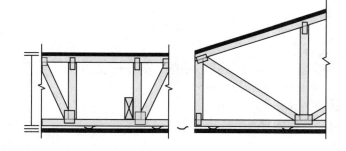

UL – L550, L521, L562, L563, L558, L574, and GA FC 5514
(floor or roof – air duct & damper, optional insulation)

Fire Rating: 1 Hr

Finish Rating: 23 Minutes

Construction: Wood trusses 24" oc, minimum 18" depth, 12" without damper

Resilient channel 16" oc, alt. 12" oc

One layer 5/8" proprietary Type X gypsum board

Sheathing minimum 23/32"

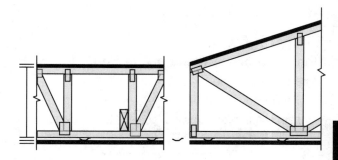

M16: FIRE DESIGN

16

UL – P522 & P531, P533, P538, P544, P545, and GA RC2603
(pitched roof – duct or damper, optional insulation)

Fire Rating: 1 Hr

Finish Rating: 25 Minutes

Construction: Wood trusses 24" oc, minimum 3" or 5-1/4" depth

Resilient channel 16" oc

One layer 5/8" proprietary Type X gypsum board

Sheathing minimum 15/32"

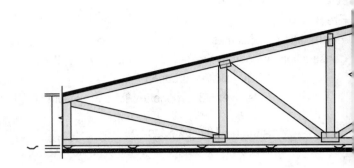

WH TSC/FCA 60-02
(floor or roof - optional insulation)

Fire Rating: 1 Hr

Finish Rating: 22 Minutes

Construction: Wood trusses (nominal 2x3) 24" oc, minimum 10" depth

Truswal metal truss plates

Resilient channel 24" oc

One layer 5/8" proprietary Type X gypsum board

Sheathing minimum 3/4"

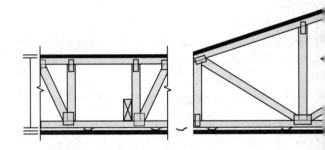

WH TSC/FCA 60-04
(floor or roof - suspended ceiling & fixtures, optional insulation)

Fire Rating: 1 Hr

Finish Rating: 27 Minutes

Construction: Wood trusses (nominal 2x3) 24" oc, minimum 10" depth

Truswal metal truss plates

Fire rated suspended ceiling - a minimum of 7-1/2" below the joist

Sheathing minimum 3/4"

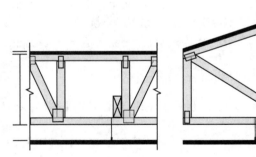

WH TSC/FCA 60-06
(floor or roof - optional insulation)

Fire Rating: 1 Hr

Finish Rating: 24 Minutes

Construction: Wood trusses (nominal 2x3) 24" oc, minimum 10" depth

Truswal metal truss plates

TrusGard Protective Channels applied to bottom chord of each truss

One layer 5/8" proprietary Type X gypsum board

Sheathing minimum 3/4"

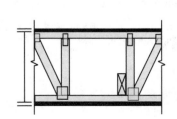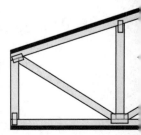

VH TSC/FCA 60-08
(floor or roof - suspended ceiling & fixtures, optional insulation)

Fire Rating: 1 Hr
Finish Rating: 29 Minutes
Construction: Wood trusses (nominal 2x3) 24" oc, minimum 10" depth

Truswal metal truss plates

TrusGard Protective Channels applied to bottom chord of each truss

Fire rated suspended ceiling - a minimum of 7-1/2" below the joist

Sheathing minimum 3/4"

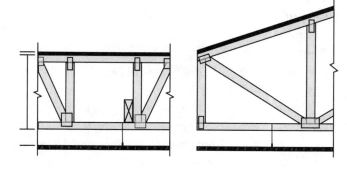

VH TSC/FCA 60-10
(floor or roof - optional insulation)

Fire Rating: 1 Hr
Finish Rating: 45 Minutes
Construction: Wood trusses (nominal 2x3) 24" oc, minimum 10" depth

Truswal metal truss plates

Two layers 1/2" Type X gypsum board

Sheathing minimum 3/4"

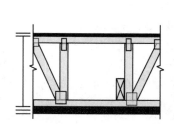
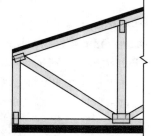

90-Minute Fire Resistive Truss Designs:

VH TSC/FCA 90-02
(floor or roof - optional insulation)

Fire Rating: 90 Minutes
Finish Rating: 45 Minutes
Construction: Wood trusses (nominal 2x3) 24" oc, minimum 10" depth

Truswal metal truss plates

Two layers 5/8" Type X gypsum board

Sheathing minimum 3/4"

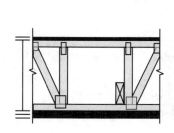
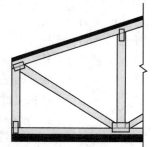

Two-Hour Fire Resistive Truss Designs:

Calculated Assembly by Kirk Grundahl, P.E., Qualtim International, 1997
(floor or roof)

Fire Rating: 2 Hr
Finish Rating: 90+ Minutes
Construction: Wood trusses (nominal 2x3) 24" oc, minimum 12" depth

Resilient channel 24" oc

Three layers 5/8" proprietary Type X gypsum board

Sheathing minimum 23/32"

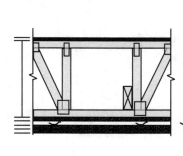
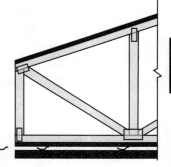

NER – 392 WTCA – FR-SYSTEM 2™
(floor or roof - optional insulation)

Fire Rating: 2 Hr

Finish Rating: 65 Minutes

Construction: Wood trusses (nominal 2x3) 24" oc, minimum 16" depth

2" shield member attached to bottom chord

FR-Quik Channel Sets™ and Bond Washers™ by Alpine Engineered Products

Note: alternate installation with resilient channel added

Two layers 5/8" proprietary Type X gypsum board

Sheathing minimum 23/32"

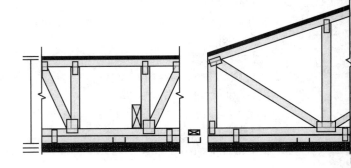

PFS 89-71, FR-SYSTEM 6™
(floor or roof - optional insulation)

Fire Rating: 2 Hr

Finish Rating: 100+ Minutes

Construction: Wood trusses 24" oc, minimum depth 9-1/4"

Resilient channel 16" oc

Three layers 5/8" proprietary Type X gypsum board

Sheathing minimum 23/32"

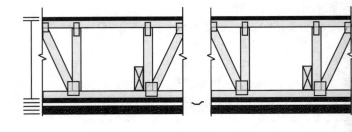

UL L-556, GA FC 5751, and RC 2751
(floor or roof – alternate truss configuration)

Fire Rating: 2 Hr

Finish Rating: 2 Hr

Construction: Wood trusses 24" oc, minimum 18" depth

Resilient channel 24" oc

Four layers 5/8" gypsum board

Sheathing 23/32"

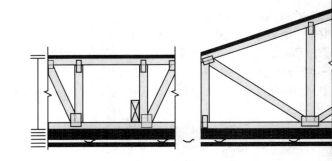

UL L577
(floor or roof - insulation)

Fire Rating: 2 Hr

Construction: Wood trusses 24" oc, minimum 12" depth

Resilient or furring channel 16" oc

Three layers 5/8" proprietary Type X gypsum board

Sheathing minimum 23/32"

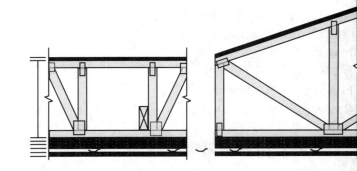

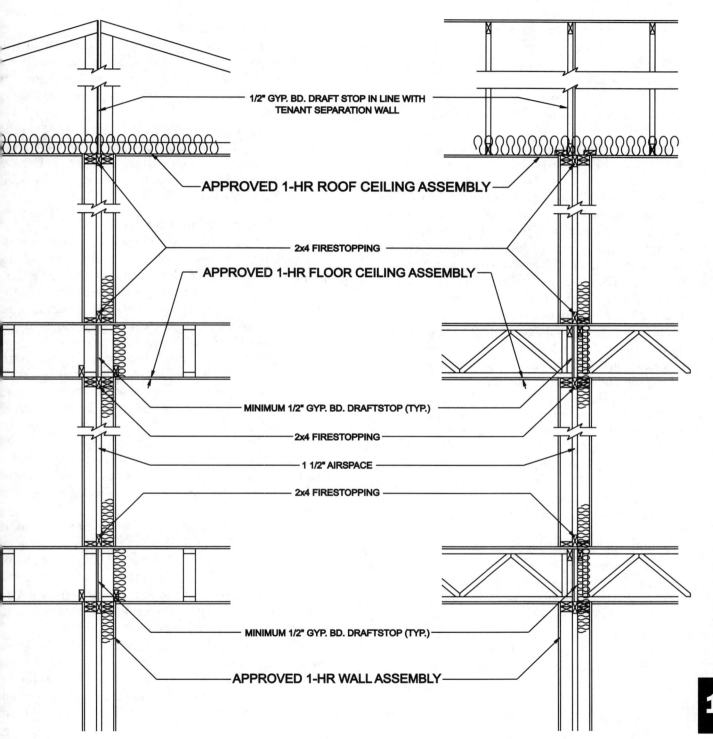

1/2" GYP. BD. DRAFT STOP IN LINE WITH
TENANT SEPARATION WALL

APPROVED 1-HR ROOF CEILING ASSEMBLY

2x4 FIRESTOPPING

APPROVED 1-HR FLOOR CEILING ASSEMBLY

MINIMUM 1/2" GYP. BD. DRAFTSTOP (TYP.)

2x4 FIRESTOPPING

1 1/2" AIRSPACE

2x4 FIRESTOPPING

MINIMUM 1/2" GYP. BD. DRAFTSTOP (TYP.)

APPROVED 1-HR WALL ASSEMBLY

M16: FIRE DESIGN

16

Area Separation Assemblies

It is of great concern when fire-rated assemblies are designed and specified without consideration of sound structural principles. Should a fire develop, these structural inadequacies could cause the assemblies to fail unexpectedly, increasing the risk of loss of life. There are a number of ways to provide sound structural and fire endurance details that maintain 1-hour rated area separation assemblies.

Figure M16.1-22 shows several possible assemblies that can be used to make up the 1-hour rated system for separation between occupancies for a) floor trusses parallel to the wall assembly; and b) perpendicular to the wall assembly. A 2x4 firestop is used between the walls next to the wall top plates. This effectively prevents the spread of fire inside the wall cavity. A minimum 1/2-inch gypsum wallboard attached to one side of the floor truss system, and located between the floor trusses, also provides a draftstop and fire protection barrier between occupancy spaces if a fire starts in the floor truss concealed space, which is a rare occurrence. The tenant separation in the roof is maintained through the use of a 1/2-inch gypsum wallboard draftstop attached to the ends of one side of the monopitch trusses and provided for the full truss height. Figure M16.1-22 effectively provides 1-hour compartmentation for all the occupied spaces using listed 1-hour rated assemblies and the appropriate draftstops for the concealed spaces as prescribed by the model building codes.

If a fire-resistive assembly, rather than draftstopping, is required within concealed attic spaces, the details shown to the right (UL U338, U339, and U377) provide approved 1-hour and 2-hour rated assemblies that may be used within the roof cavity and that may be constructed with gable end frames.

The critical aspects for fire endurance assemblies include:

- Ensuring that the wall and ceiling assemblies of the room use 1-hour rated assemblies. These are independent assemblies. The wall assembly does not have to be continuous from the floor to the roof to meet the intent of the code or the fire endurance performance of the structure. The intent of the code is that the building be broken into compartments to contain a fire to a given area. Fire resistance assemblies are tested to provide code-complying fire endurance to meet the intent of the code. The foregoing details meet the intent of the code.
- Properly fastening the gypsum wallboard to the wall studs and trusses. This is critical for achieving the desired fire performance from a UL or GA assembly.

- Ensuring that the detail being used is structurally sound, particularly the bearing details. All connection details are critical to assembly performance. When a fire begins, if the structural detail is poor, the system will fail at the poor connection detail earlier than expected.
- Accommodating both sound structural details with appropriate fire endurance details. Since all conceivable field conditions have not been and cannot be tested, rational engineering judgment needs to be used.

The foregoing principles could also be applied to structures requiring 2-hour rated area separation assemblies. In this case, acceptable 2-hour wall assemblies would be used in conjunction with the 2-hour floor-ceiling and roof-ceiling fire endurance assemblies.

Through-Penetration Fire Stops

Because walls and floors are penetrated for a variety of plumbing, ventilation, electrical, and communication purposes, ASTM E814 *Standard Method of Fire Tests of Through-Penetration Fire Stops* uses fire-resistive assemblies (rated per ASTM E119) and penetrates them with cables, pipes, and ducts, etc., before subjecting the assembly to ASTM E119's fire endurance tests.

Despite the penetrations, firestop assemblies tested according to ASTM E814 must not significantly lose their fire containment properties in order to be considered acceptable. Properties are measured according to the passage of any heat, flame, hot gases, or combustion through the firestop to the test assembly's unexposed surface.

Upon successful completion of fire endurance test, firestop systems are given F&T ratings. These ratings are expressed in hourly terms, in much the same fashion as fire-resistive barriers.

To obtain an F-Rating, a firestop must remain in the opening during the fire and hose stream test, withstanding the fire test for a prescribed rating period without permitting the passage of flame on any element of its unexposed side. During the hose stream test, a firestop must not develop any opening that would permit a projection of water from the stream beyond the unexposed side.

To obtain a T-Rating, a firestop must meet the requirements of the F-Rating. In addition, the firestop must prevent the transmission of heat during the prescribed rating period which would increase the temperature of any thermocouple on its unexposed surface, or any penetrating items, by more than 325°F.

The UL Fire Resistance Directory lists literally thousands of tested systems. They have organized the systems with an alpha-alpha-numeric identification number. The first alpha is either a F, W, or C. These letters signify the

UL – U338
(wall – bearing/non-bearing, optional insulation)
Fire Rating: 1 Hr

Finish Rating: 20 Minutes with one layer, 59 Minutes with two layers gypsum

Construction: Wood gable truss

2 by 3 or 2 by 4 studs spaced 24" oc max, effectively firestopped

One layer 5/8" gypsum board each side for non-bearing wall

Two layers 5/8" gypsum board each side for bearing wall

UL – U339
(wall – bearing/non-bearing, optional insulation)
Fire Rating: 1 Hr

Finish Rating: 20 Minutes with one layer, 59 Minutes with two layers gypsum

Construction: 2 Wood gable trusses

2 by 3 or 2 by 4 studs spaced 24" oc max, effectively firestopped

One layer 5/8" gypsum board each side for non-bearing wall

Two layers 5/8" gypsum board each side for bearing wall

Septum sheathed with plywood or mineral/fiber board (optional in bearing configuration)

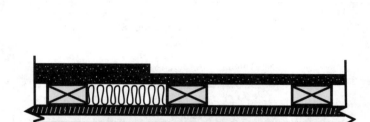

UL – U377
(wall – bearing, required insulation)
Fire Rating: 2 Hr

Finish Rating: 47 Minutes with two layers gypsum

Construction: Double row of nominal 2x4 studs, flat-wise, spaced 24" oc max, effectively firestopped

Two layers 5/8" Type X gypsum board each side

Septum filled with spray applied cellulose material

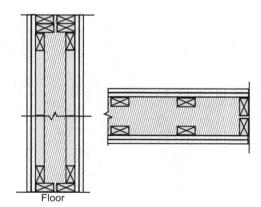

Floor

type of assembly being penetrated: F signifies a floor, W signifies a wall, and C signifies a ceiling. The second alpha signifies a limiting description, for example: C signifies a framed floor, and L signifies a framed wall. The numeric portion is also significant: 0000-0999 signifies no penetrating items, 1000-1999 signifies metallic pipe, 2000-2999 signifies nonmetallic pipe, 3000-3999 signifies electrical cable, 4000-4999 signifies cable trays, 5000-5999 signifies insulated pipes, 6000-6999 signifies miscellaneous electrical penetrants, 7000-7999 signifies miscellaneous mechanical penetrants, and 8000-8999 signifies a combination of penetrants.

For example, a floor/ceiling penetrated by a metallic pipe would have a designation of F-C-1xxx, or with nonmetallic pipe would be designated F-C-2xxx. A few examples of floor/ceiling penetrations are included below. Examples in Figure M16.1-23 are from the UL Fire Resistance Directory, Vol. II: UL Systems F-C-2008, F-C-1006, F-C-3007 & 8, F-C-5002.

Figure M16.1-23 Examples of Through-Penetration Firestop Systems

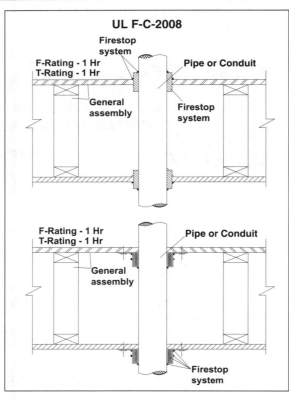

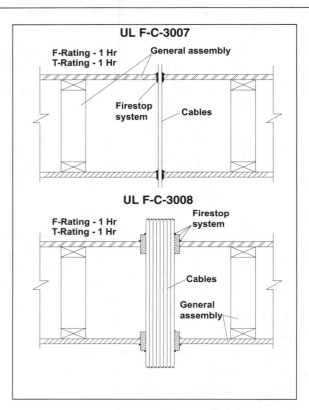

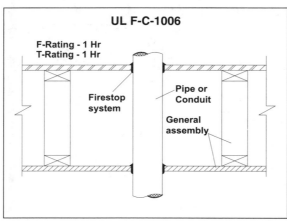

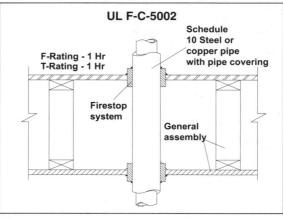

Transitory Floor Vibration and Sound Transmission

Sound Transmission

Sound transmission ratings are closely aligned with fire endurance ratings for assemblies. This is due to the fact that flame and sound penetrations follow similar paths of least resistance.

Sound striking a wall or ceiling surface is transmitted through the air in the wall or ceiling cavity. It then strikes the opposite wall surface, causing it to vibrate and transmit the sound into the adjoining room. Sound also is transmitted through any openings into the room, such as air ducts, electrical outlets, window openings, and doors. This is airborne sound transmission. The Sound Transmission Class (STC) method of rating airborne sounds evaluates the comfort ability of a particular living space. The higher the STC, the better the airborne noise control performance of the structure. An STC of 50 or above is generally considered a good airborne noise control rating. Table M16.1-6 describes the privacy afforded according to the STC rating.

Impact Sound Transmission is produced when a structural element is set into vibration by direct impact – someone walking, for example. The vibrating surface generates sound waves on both sides of the element. The Impact Insulation Class (IIC) is a method of rating the impact sound transmission performance of an assembly. The higher the IIC, the better the impact noise control of the element. An IIC of 55 is generally considered a good impact noise control.

Estimated Wood Floor Sound Performance[1,2,3]

Sound transmission and impact insulation characteristics of a floor assembly can be calculated in a manner similar to fire calculations – by adding up the value of the individual components. The contributions of various products to an STC or IIC rating are shown in Table M16.1-9. An example calculation is shown in Table M16.1-10.

Tables M16.1-11 and M16.1-12 provide STC and IIC ratings for specific 1-hour fire-resistive metal plate connected wood truss assemblies. Ratings are provided for various floor coverings including combinations of carpet and pad, vinyl, lightweight concrete, and gypcrete.

Table M16.1-8 Privacy Afforded According to STC Rating

STC Rating	Privacy Afforded
25	Normal speech easily understood
30	Normal speech audible, but not intelligible
35	Loud speech audible and fairly understandable
40	Loud speech barely audible, but not intelligible
45	Loud speech barely audible
50	Shouting barely audible
55	Shouting inaudible

Table M16.1-9 Contributions of Various Products to STC or IIC Rating

Description	Frequency	
	STC High	IIC Low
Basic wood floor - consisting of wood joist (I-joist, solid sawn, or truss) 3/4" decking and 5/8" gypsum wallboard directly attached to ceiling	36	33
Cushioned vinyl or linoleum	0	2
Non-cushioned vinyl or linoleum	0	0
1/2" parquet flooring	0	1
3/4" Gypcrete® or Elastizel®	7 to 8	1
1 1/2" lightweight concrete	7 to 8	1
1/2" sound deadening board (USG)[1]	1	5
Quiet-Cor® underlayment by Tarkett, Inc.[1]	1	8
Enkasonic® by American Enka Company[1]	4	13
Sempafloor® by Laminating Services, Inc.[1]	1	11
R-19 batt insulation	2	0
R-11 batt insulation	1	0
3" mineral wool insulation	1	0
Resilient channel	10	8
Resilient with insulation	13	15
Extra layer of 5/8" gypsum wallboard	0 to 2	2 to 4
Carpet & pad	0	20 to 25

1. Estimates based on proprietary literature. Verify with individual companies.

M16: FIRE DESIGN

16

1. *Acoustical Manual*, National Association of Home Builders, 1978.
2. Yerges, Lyle F., *Sound, Noise and Vibration Control*, 1969.
3. *Catalog of STC and IIC Ratings for Wall and Floor/Ceiling Assemblies*, California Dept. of Health Services, Office of Noise Control, Berkeley, CA.

Table M16.1-10 Example Calculation

Description	STC	IIC
Carpet & pad	0	20
3/4" Gypcrete	7	1
Wood I-joist floor	36	33
Resilient channel	10	8
Total	*53*	*62*

Table M16.1-11 STC & IIC Ratings for UL L528/L529

Floor Covering	STC	IIC	Test Number
Carpet & Pad	48	56	NRC 1039 & 1040
Vinyl	45	37	NRC 1041 & 1042
Lightweight, Carpet & Pad	57	72	NRC 1044 & 1045
Lightweight and Vinyl	57	50*	NRC 1047 & 1048
Gypcrete & Cushioned Vinyl	--	53	6-442-2 Gypcrete
Gypcrete, Carpet & Pad	--	74	6-442-3 Gypcrete
Gypcrete	58	--	6-442-5 Gypcrete

* Does not match source document which was in error.

Table M16.1-12 STC & IIC Ratings for FC-214

Floor Covering	STC	IIC	Test Number
Carpet & Pad	48	54	NRC 1059 & 1060
Vinyl	47	35	NRC 1063 & 1064
Lightweight, Carpet & Pad	56	72	NRC 1053 & 1054
Lightweight and Vinyl	56	48	NRC 1051 & 1052
Gypcrete, Carpet & Pad	52	63	NRC 1076 & 1077
Gypcrete	53	43	NRC 1085 & 1086

Description of Materials:

Gypcrete	3/4"
Lightweight Concrete	1"
Carpet	2.63 Kg/M^2
Pad	1.37 Kg/M^2

Tables used with permission
of Truss Plate Institute, Inc.

M16.2 Design Procedures for Exposed Wood Members

For members stressed in one principle direction, simplifications can be made which allow the tabulation of load factor tables for fire design. These load factor tables can be used to determine the structural design load ratio, R_s, at which the member has sufficient capacity for a given fire endurance time. This section provides the rational used to develop the load ratio tables provided later in this section (Tables M16.2-1 through M16.2-10). For more complex calculations where stress interactions must be considered, use the provisions of AF&PA's *Technical Report 10* with the appropriate *NDS* provisions.

Bending Members (Tables M16.2-1 through M16.2-2)

Structural: $D+L \leq R_s F_b S_s C_{L\text{-}s} C_D C_M C_t$

Fire: $D+L \leq 2.85 F_b S_f C_{L\text{-}f}$

where:

D = Design dead load

L = Design live load

R_s = Design load ratio

F_b = Tabulated bending design value

S_s = Section modulus using full cross-section dimensions

S_f = Section modulus using cross-section dimensions reduced from fire exposure

$C_{L\text{-}s}$ = Beam stability factor using full cross-section dimensions

$C_{L\text{-}f}$ = Beam stability factor using cross-section dimensions reduced from fire exposure

C_D = Load duration factor

C_M = Wet service factor

C_t = Temperature factor

Solve for R_s:

$$R_s = \frac{2.85\ S_f\ C_{L\text{-}f}}{S_s\ C_{L\text{-}s}\ C_D\ C_M\ C_t} \qquad \text{(M16.2-1)}$$

Load ratio tables were developed for standard reference conditions where: $C_D = 1.0$; $C_M = 1.0$; $C_t = 1.0$; $C_{L\text{-}f} = 1.0$

The calculation of $C_{L\text{-}s}$ and $C_{L\text{-}f}$ require the designer to consider both the change in bending section relative to bending strength and the change in buckling stiffness relative to buckling strength. While these relationships can be directly calculated using *NDS* provisions, they can not be easily tabulated. However, for most beams exposed on three sides, the beams are braced on the protected side. For long span beams exposed on four sides, the beam failure is influenced by buckling due to lateral instability. When buckling is considered, the following equations should be used:

Structural (buckling): $D+L \leq R_s E_{min} I_{yy\text{-}s} / \ell_e C_M C_t$

Fire (buckling): $D+L \leq 2.03 E_{min} I_{yy\text{-}f} / \ell_e$

where:

D = Design dead load

L = Design live load

R_s = Design load ratio (buckling)

E_{min} = Reference modulus of elasticity for beam stability calculations

$I_{yy\text{-}s}$ = Lateral moment of inertia using full cross-section dimensions

$I_{yy\text{-}f}$ = Lateral moment of inertia using cross-section dimensions reduced from fire exposure

C_M = Wet service factor

C_t = Temperature factor

$$R_s = \frac{2.03\ I_{yy\text{-}f}}{I_{yy\text{-}s}\ C_M\ C_t} \qquad \text{(M16.2-2)}$$

Compression Members (Tables M16.2-3 through M16.2-5)

Structural: $D+L \leq R_s F_c C_{p-s} C_D C_M C_t$

Fire: $D+L \leq 2.58 F_c C_{p-f}$

where:

D = Design dead load

L = Design live load

R_s = Design load ratio

F_c = Tabulated compression parallel-to-grain design value

C_{p-s} = Column stability factor using full cross-section dimensions

C_{p-f} = Column stability factor using cross-section dimensions reduced from fire exposure

C_D = Load duration factor

C_M = Wet service factor

C_t = Temperature factor

The calculation of C_{p-s} and C_{p-f} require the designer to consider both the change in compression area relative to compression parallel-to-grain strength and the change in buckling stiffness relative to buckling strength. While these relationships can be directly calculated using *NDS* provisions, they can not be easily tabulated. However, for most column fire endurance designs the mode of column failure is significantly influenced by buckling. For this reason, conservative load ratio tables can be tabulated for changes in buckling capacity as a function of fire exposure.

Structural (buckling): $D+L \leq R_s \pi^2 E_{min} I_s / \ell_e^2 C_M C_t$

Fire (buckling): $D+L \leq 2.03 \pi^2 E_{min} I_f / \ell_e^2$

where:

D = Design dead load

L = Design live load

R_s = Design load ratio (buckling)

E_{min} = Reference modulus of elasticity for column stability calculations

I_s = Moment of inertia using full cross-section dimensions

I_f = Moment of inertia using cross-section dimensions reduced from fire exposure

C_M = Wet service factor

C_t = Temperature factor

$$R_s = \frac{2.03 \ I_f}{I_s \ C_M \ C_t} \qquad \text{(M16.2-3}$$

Buckling load ratio tables were developed for standar reference conditions where: $C_M = 1.0$; $C_t = 1.0$

NOTE: The load duration factor, C_D, is not included in the load ratio tables since modulus of elasticity values, E, used in the buckling capacity calculation is not adjusted for load duration in the *NDS*.

Tension Members (Tables M16.2-6 through M16.2-8)

Structural: $D+L \leq R_s F_t A_s C_D C_M C_t C_i$

Fire: $D+L \leq 2.85 F_t A_f$

where:

D = Design dead load

L = Design live load

R_s = Design load ratio

F_t = Tabulated tension parallel-to-grain design value

A_s = Area of cross section using full cross-section dimensions

A_f = Area of cross section using cross-section dimensions reduced from fire exposure

C_D = Load duration factor

C_M = Wet service factor

C_t = Temperature factor

$$R_s = \frac{2.85 \ A_f}{A_s \ C_D \ C_M \ C_t} \qquad \text{(M16.2-4}$$

Load ratio tables were developed for standard refe ence conditions where: $C_D = 1.0$; $C_M = 1.0$; $C_t = 1.0$

Table M16.2-1 Design Load Ratios for Bending Members Exposed on Three Sides

(Structural Calculations at Standard Reference Conditions: $C_D = 1.0$, $C_M = 1.0$, $C_t = 1.0$, $C_i = 1.0$, $C_L = 1.0$)
(Protected Surface in Depth Direction)

Table M16.2-1A Southern Pine Structural Glued Laminated Timbers

Rating	1-HOUR				1.5-HOUR			2-HOUR	
Beam Width	5	6.75	8.5	10.5	6.75	8.5	10.5	8.5	10.5
Beam Depth	Design Load Ratio, R_s								
5.5	0.36	0.60	0.74	0.85	0.22	0.35	0.44	0.13	0.20
6.875	0.43	0.72	0.90	1.00	0.30	0.47	0.60	0.21	0.33
8.25	0.49	0.81	1.00	1.00	0.36	0.57	0.72	0.28	0.43
9.625	0.53	0.88	1.00	1.00	0.40	0.64	0.82	0.33	0.51
11	0.56	0.93	1.00	1.00	0.44	0.70	0.89	0.37	0.58
12.375	0.58	0.97	1.00	1.00	0.47	0.75	0.95	0.40	0.63
13.75	0.60	1.00	1.00	1.00	0.49	0.78	1.00	0.43	0.67
15.125	0.62	1.00	1.00	1.00	0.51	0.82	1.00	0.46	0.71
16.5	0.63	1.00	1.00	1.00	0.53	0.84	1.00	0.48	0.74
17.875	0.65	1.00	1.00	1.00	0.54	0.87	1.00	0.49	0.77
19.25	0.66	1.00	1.00	1.00	0.56	0.89	1.00	0.51	0.79
20.625	0.66	1.00	1.00	1.00	0.57	0.90	1.00	0.52	0.81
22	0.67	1.00	1.00	1.00	0.58	0.92	1.00	0.53	0.83
23.375	0.68	1.00	1.00	1.00	0.59	0.93	1.00	0.55	0.85
24.75	0.69	1.00	1.00	1.00	0.60	0.95	1.00	0.55	0.86
26.125	0.69	1.00	1.00	1.00	0.60	0.96	1.00	0.56	0.88
27.5	0.70	1.00	1.00	1.00	0.61	0.97	1.00	0.57	0.89
28.875	0.70	1.00	1.00	1.00	0.61	0.98	1.00	0.58	0.90
30.25	0.71	1.00	1.00	1.00	0.62	0.99	1.00	0.58	0.91
31.625	0.71	1.00	1.00	1.00	0.62	0.99	1.00	0.59	0.92
33	0.71	1.00	1.00	1.00	0.63	1.00	1.00	0.60	0.93
34.375	0.72	1.00	1.00	1.00	0.63	1.00	1.00	0.60	0.93
35.75	0.72	1.00	1.00	1.00	0.64	1.00	1.00	0.61	0.94
37.125		1.00	1.00	1.00	0.64	1.00	1.00	0.61	0.95
38.5		1.00	1.00	1.00	0.64	1.00	1.00	0.61	0.95
39.875		1.00	1.00	1.00	0.65	1.00	1.00	0.62	0.96
41.25		1.00	1.00	1.00	0.65	1.00	1.00	0.62	0.97
42.625		1.00	1.00	1.00	0.65	1.00	1.00	0.63	0.97
44		1.00	1.00	1.00	0.66	1.00	1.00	0.63	0.98
45.375		1.00	1.00	1.00	0.66	1.00	1.00	0.63	0.98
46.75		1.00	1.00	1.00	0.66	1.00	1.00	0.63	0.99
48.125		1.00	1.00	1.00	0.66	1.00	1.00	0.64	0.99
49.5			1.00	1.00		1.00	1.00	0.64	0.99
50.875			1.00	1.00		1.00	1.00	0.64	1.00
52.25			1.00	1.00		1.00	1.00	0.64	1.00
53.625			1.00	1.00		1.00	1.00	0.65	1.00
55			1.00	1.00		1.00	1.00	0.65	1.00
56.375			1.00	1.00		1.00	1.00	0.65	1.00
57.75			1.00	1.00		1.00	1.00	0.65	1.00
59.125			1.00	1.00		1.00	1.00	0.65	1.00
60.5			1.00	1.00		1.00	1.00	0.66	1.00
61.875			1.00	1.00		1.00	1.00	0.66	1.00
63.25			1.00	1.00		1.00	1.00	0.66	1.00
64.625				1.00			1.00		1.00
66				1.00			1.00		1.00
67.375				1.00			1.00		1.00
68.75				1.00			1.00		1.00
70.125				1.00			1.00		1.00
71.5				1.00			1.00		1.00
72.875				1.00			1.00		1.00
74.25				1.00			1.00		1.00
75.625				1.00			1.00		1.00
77				1.00			1.00		1.00

Table M16.2-1B Western Species Structural Glued Laminated Timbers

Rating	1-HOUR				1.5-HOUR			2-HOUR	
Beam Width	5.125	6.75	8.75	10.75	6.75	8.75	10.75	8.75	10.75
Beam Depth	Design Load Ratio, R_s								
6	0.42	0.65	0.82	0.93	0.25	0.41	0.52	0.18	0.26
7.5	0.49	0.77	0.97	1.00	0.33	0.54	0.68	0.26	0.39
9	0.54	0.85	1.00	1.00	0.38	0.64	0.79	0.33	0.49
10.5	0.58	0.91	1.00	1.00	0.43	0.71	0.88	0.39	0.57
12	0.61	0.96	1.00	1.00	0.46	0.76	0.95	0.43	0.64
13.5	0.64	1.00	1.00	1.00	0.49	0.81	1.00	0.46	0.69
15	0.66	1.00	1.00	1.00	0.51	0.85	1.00	0.49	0.73
16.5	0.67	1.00	1.00	1.00	0.53	0.88	1.00	0.52	0.77
18	0.69	1.00	1.00	1.00	0.55	0.90	1.00	0.54	0.80
19.5	0.70	1.00	1.00	1.00	0.56	0.93	1.00	0.55	0.82
21	0.71	1.00	1.00	1.00	0.57	0.95	1.00	0.57	0.85
22.5	0.72	1.00	1.00	1.00	0.58	0.96	1.00	0.58	0.87
24	0.73	1.00	1.00	1.00	0.59	0.98	1.00	0.60	0.88
25.5	0.73	1.00	1.00	1.00	0.60	0.99	1.00	0.61	0.90
27	0.74	1.00	1.00	1.00	0.61	1.00	1.00	0.62	0.91
28.5	0.74	1.00	1.00	1.00	0.61	1.00	1.00	0.62	0.93
30	0.75	1.00	1.00	1.00	0.62	1.00	1.00	0.63	0.94
31.5	0.75	1.00	1.00	1.00	0.62	1.00	1.00	0.64	0.95
33	0.76	1.00	1.00	1.00	0.63	1.00	1.00	0.65	0.96
34.5	0.76	1.00	1.00	1.00	0.63	1.00	1.00	0.65	0.97
36	0.77	1.00	1.00	1.00	0.64	1.00	1.00	0.66	0.98
37.5		1.00	1.00	1.00	0.64	1.00	1.00	0.66	0.98
39		1.00	1.00	1.00	0.64	1.00	1.00	0.67	0.99
40.5		1.00	1.00	1.00	0.65	1.00	1.00	0.67	1.00
42		1.00	1.00	1.00	0.65	1.00	1.00	0.68	1.00
43.5		1.00	1.00	1.00	0.65	1.00	1.00	0.68	1.00
45		1.00	1.00	1.00	0.66	1.00	1.00	0.68	1.00
46.5		1.00	1.00	1.00	0.66	1.00	1.00	0.69	1.00
48		1.00	1.00	1.00	0.66	1.00	1.00	0.69	1.00
49.5			1.00	1.00		1.00	1.00	0.69	1.00
51			1.00	1.00		1.00	1.00	0.70	1.00
52.5			1.00	1.00		1.00	1.00	0.70	1.00
54			1.00	1.00		1.00	1.00	0.70	1.00
55.5			1.00	1.00		1.00	1.00	0.70	1.00
57			1.00	1.00		1.00	1.00	0.70	1.00
58.5			1.00	1.00		1.00	1.00	0.71	1.00
60			1.00	1.00		1.00	1.00	0.71	1.00
61.5			1.00	1.00		1.00	1.00	0.71	1.00
63			1.00	1.00		1.00	1.00	0.71	1.00
64.5				1.00			1.00		1.00
66				1.00			1.00		1.00
67.5				1.00			1.00		1.00
69				1.00			1.00		1.00
70.5				1.00			1.00		1.00
72				1.00			1.00		1.00
73.5				1.00			1.00		1.00
75				1.00			1.00		1.00
76.5				1.00			1.00		1.00
78				1.00			1.00		1.00
79.5				1.00			1.00		1.00
81				1.00			1.00		1.00

Note: Tabulated values assume bending in the depth direction.

Table M16.2-1C Solid Sawn Timbers

Rating	1-HOUR				1.5-HOUR			2-HOUR	
Beam Width	5.5	7.5	9.5	11.5	7.5	9.5	11.5	9.5	11.5
Beam Depth	Design Load Ratio, R_s								
5.5	0.45	0.67	0.80	0.89	0.28	0.40	0.48	0.17	0.23
7.5	0.57	0.86	1.00	1.00	0.42	0.60	0.71	0.32	0.43
9.5	0.65	0.97	1.00	1.00	0.51	0.73	0.87	0.42	0.57
11.5	0.70	1.00	1.00	1.00	0.58	0.83	0.99	0.50	0.67
13.5	0.74	1.00	1.00	1.00	0.63	0.89	1.00	0.56	0.75
15.5	0.77	1.00	1.00	1.00	0.67	0.95	1.00	0.60	0.81
17.5	0.79	1.00	1.00	1.00	0.70	0.99	1.00	0.64	0.86
19.5	0.81	1.00	1.00	1.00	0.72	1.00	1.00	0.67	0.90
21.5	0.83	1.00	1.00	1.00	0.74	1.00	1.00	0.69	0.93
23.5	0.84	1.00	1.00	1.00	0.76	1.00	1.00	0.71	0.96

M16: FIRE DESIGN

16

Table M16.2-2 Design Load Ratios for Bending Members Exposed on Four Sides

(Structural Calculations at Standard Reference Conditions: $C_D = 1.0$, $C_M = 1.0$, $C_t = 1.0$, $C_i = 1.0$, $C_L = 1.0$)

Table M16.2-2A Southern Pine Structural Glued Laminated Timbers

Rating	1-HOUR				1.5-HOUR			2-HOUR	
Beam Width	5	6.75	8.5	10.5	6.75	8.5	10.5	8.5	10.5
Beam Depth	Design Load Ratio, R_s								
5.5	0.10	0.16	0.20	0.22	0.01	0.01	0.01	0.02	0.03
6.875	0.18	0.30	0.37	0.42	0.05	0.09	0.11	0.00	0.01
8.25	0.25	0.42	0.52	0.59	0.11	0.18	0.23	0.04	0.06
9.625	0.31	0.52	0.64	0.73	0.17	0.27	0.34	0.09	0.13
11	0.36	0.60	0.74	0.85	0.22	0.35	0.44	0.13	0.20
12.375	0.40	0.67	0.83	0.94	0.26	0.42	0.53	0.17	0.27
13.75	0.43	0.72	0.90	1.00	0.30	0.47	0.60	0.21	0.33
15.125	0.46	0.77	0.95	1.00	0.33	0.52	0.67	0.25	0.38
16.5	0.49	0.81	1.00	1.00	0.36	0.57	0.72	0.28	0.43
17.875	0.51	0.85	1.00	1.00	0.38	0.61	0.77	0.30	0.47
19.25	0.53	0.88	1.00	1.00	0.40	0.64	0.82	0.33	0.51
20.625	0.54	0.91	1.00	1.00	0.42	0.67	0.86	0.35	0.54
22	0.56	0.93	1.00	1.00	0.44	0.70	0.89	0.37	0.58
23.375	0.57	0.95	1.00	1.00	0.45	0.72	0.92	0.39	0.60
24.75	0.58	0.97	1.00	1.00	0.47	0.75	0.95	0.40	0.63
26.125	0.59	0.99	1.00	1.00	0.48	0.77	0.97	0.42	0.65
27.5	0.60	1.00	1.00	1.00	0.49	0.78	1.00	0.43	0.67
28.875	0.61	1.00	1.00	1.00	0.50	0.80	1.00	0.44	0.69
30.25	0.62	1.00	1.00	1.00	0.51	0.82	1.00	0.46	0.71
31.625	0.63	1.00	1.00	1.00	0.52	0.83	1.00	0.47	0.73
33	0.63	1.00	1.00	1.00	0.53	0.84	1.00	0.48	0.74
34.375	0.64	1.00	1.00	1.00	0.54	0.86	1.00	0.49	0.75
35.75	0.65	1.00	1.00	1.00	0.54	0.87	1.00	0.49	0.77
37.125		1.00	1.00	1.00	0.55	0.88	1.00	0.50	0.78
38.5		1.00	1.00	1.00	0.56	0.89	1.00	0.51	0.79
39.875		1.00	1.00	1.00	0.56	0.90	1.00	0.52	0.80
41.25		1.00	1.00	1.00	0.57	0.90	1.00	0.52	0.81
42.625		1.00	1.00	1.00	0.57	0.91	1.00	0.53	0.82
44		1.00	1.00	1.00	0.58	0.92	1.00	0.53	0.83
45.375		1.00	1.00	1.00	0.58	0.93	1.00	0.54	0.84
46.75		1.00	1.00	1.00	0.59	0.93	1.00	0.55	0.85
48.125		1.00	1.00	1.00	0.59	0.94	1.00	0.55	0.85
49.5			1.00	1.00		0.95	1.00	0.55	0.86
50.875			1.00	1.00		0.95	1.00	0.56	0.87
52.25			1.00	1.00		0.96	1.00	0.56	0.88
53.625			1.00	1.00		0.96	1.00	0.57	0.88
55			1.00	1.00		0.97	1.00	0.57	0.89
56.375			1.00	1.00		0.97	1.00	0.57	0.89
57.75			1.00	1.00		0.98	1.00	0.58	0.90
59.125			1.00	1.00		0.98	1.00	0.58	0.90
60.5			1.00	1.00		0.99	1.00	0.58	0.91
61.875			1.00	1.00		0.99	1.00	0.59	0.91
63.25			1.00	1.00		0.99	1.00	0.59	0.92
64.625				1.00			1.00		0.92
66				1.00			1.00		0.93
67.375				1.00			1.00		0.93
68.75				1.00			1.00		0.93
70.125				1.00			1.00		0.94
71.5				1.00			1.00		0.94
72.875				1.00			1.00		0.95
74.25				1.00			1.00		0.95
75.625				1.00			1.00		0.95
77				1.00			1.00		0.95

Table M16.2-2B Western Species Structural Glued Laminated Timbers

Rating	1-HOUR				1.5-HOUR			2-HOUR	
Beam Width	5.125	6.75	8.75	10.75	6.75	8.75	10.75	8.75	10.75
Beam Depth	Design Load Ratio, R_s								
6	0.14	0.21	0.27	0.30	0.02	0.03	0.04	0.00	0.00
7.5	0.23	0.36	0.45	0.51	0.08	0.13	0.17	0.02	0.03
9	0.31	0.48	0.60	0.68	0.15	0.24	0.30	0.07	0.10
10.5	0.37	0.57	0.72	0.82	0.20	0.33	0.42	0.12	0.19
12	0.42	0.65	0.82	0.93	0.25	0.41	0.52	0.18	0.26
13.5	0.46	0.72	0.90	1.00	0.29	0.48	0.60	0.22	0.33
15	0.49	0.77	0.97	1.00	0.33	0.54	0.68	0.26	0.39
16.5	0.52	0.81	1.00	1.00	0.36	0.59	0.74	0.30	0.45
18	0.54	0.85	1.00	1.00	0.38	0.64	0.79	0.33	0.49
19.5	0.56	0.88	1.00	1.00	0.41	0.67	0.84	0.36	0.54
21	0.58	0.91	1.00	1.00	0.43	0.71	0.88	0.39	0.57
22.5	0.60	0.94	1.00	1.00	0.45	0.74	0.92	0.41	0.61
24	0.61	0.96	1.00	1.00	0.46	0.76	0.95	0.43	0.64
25.5	0.63	0.98	1.00	1.00	0.48	0.79	0.98	0.45	0.66
27	0.64	1.00	1.00	1.00	0.49	0.81	1.00	0.46	0.69
28.5	0.65	1.00	1.00	1.00	0.50	0.83	1.00	0.48	0.71
30	0.66	1.00	1.00	1.00	0.51	0.85	1.00	0.49	0.73
31.5	0.67	1.00	1.00	1.00	0.52	0.86	1.00	0.50	0.75
33	0.67	1.00	1.00	1.00	0.53	0.88	1.00	0.52	0.77
34.5	0.68	1.00	1.00	1.00	0.54	0.89	1.00	0.53	0.78
36	0.69	1.00	1.00	1.00	0.55	0.90	1.00	0.54	0.80
37.5		1.00	1.00	1.00	0.55	0.92	1.00	0.55	0.81
39		1.00	1.00	1.00	0.56	0.93	1.00	0.55	0.82
40.5		1.00	1.00	1.00	0.57	0.94	1.00	0.56	0.84
42		1.00	1.00	1.00	0.57	0.95	1.00	0.57	0.85
43.5		1.00	1.00	1.00	0.58	0.96	1.00	0.58	0.86
45		1.00	1.00	1.00	0.58	0.96	1.00	0.58	0.87
46.5		1.00	1.00	1.00	0.59	0.97	1.00	0.59	0.88
48		1.00	1.00	1.00	0.59	0.98	1.00	0.60	0.88
49.5			1.00	1.00		0.99	1.00	0.60	0.89
51			1.00	1.00		0.99	1.00	0.61	0.90
52.5			1.00	1.00		1.00	1.00	0.61	0.91
54			1.00	1.00		1.00	1.00	0.62	0.91
55.5			1.00	1.00		1.00	1.00	0.62	0.92
57			1.00	1.00		1.00	1.00	0.62	0.93
58.5			1.00	1.00		1.00	1.00	0.63	0.93
60			1.00	1.00		1.00	1.00	0.63	0.94
61.5			1.00	1.00		1.00	1.00	0.64	0.94
63			1.00	1.00		1.00	1.00	0.64	0.95
64.5			1.00			1.00			0.95
66			1.00			1.00			0.96
67.5			1.00			1.00			0.96
69			1.00			1.00			0.97
70.5			1.00			1.00			0.97
72			1.00			1.00			0.98
73.5			1.00			1.00			0.98
75			1.00			1.00			0.98
76.5			1.00			1.00			0.99
78			1.00			1.00			0.99
79.5			1.00			1.00			0.99
81			1.00			1.00			1.00

Note: Tabulated values assume bending in the depth direction.

Table M16.2-2C Solid Sawn Timbers

Rating	1-HOUR				1.5-HOUR			2-HOUR	
Beam Width	5.5	7.5	9.5	11.5	7.5	9.5	11.5	9.5	11.5
Beam Depth	Design Load Ratio, R_s								
5.5	0.12	0.18	0.21	0.23	0.01	0.01	0.01	0.02	0.03
7.5	0.27	0.40	0.48	0.53	0.10	0.15	0.18	0.02	0.03
9.5	0.38	0.57	0.68	0.76	0.21	0.30	0.36	0.11	0.14
11.5	0.46	0.70	0.84	0.92	0.30	0.43	0.51	0.19	0.26
13.5	0.53	0.80	0.95	1.00	0.38	0.53	0.64	0.27	0.36
15.5	0.58	0.87	1.00	1.00	0.43	0.62	0.74	0.33	0.45
17.5	0.62	0.93	1.00	1.00	0.48	0.69	0.82	0.39	0.52
19.5	0.65	0.99	1.00	1.00	0.52	0.74	0.89	0.43	0.59
21.5	0.68	1.00	1.00	1.00	0.56	0.79	0.95	0.47	0.64
23.5	0.71	1.00	1.00	1.00	0.59	0.84	1.00	0.51	0.69

Table M16.2-3 Design Load Ratios for Compression Members Exposed on Three Sides

(Structural Calculations at Standard Reference Conditions: $C_M = 1.0$, $C_t = 1.0$, $C_i = 1.0$)

(Protected Surface in Depth Direction)

Table M16.2-3A Southern Pine Structural Glued Laminated Timbers

Rating	1-HOUR				1.5-HOUR			2-HOUR	
Beam Width	5	6.75	8.5	10.5	6.75	8.5	10.5	8.5	10.5
Beam Depth	Design Load Ratio, R_s								
5.5	0.03								
6.875	0.03	0.15			0.02				
8.25	0.03	0.16	0.30		0.02	0.10		0.02	
9.625	0.04	0.17	0.32	0.47	0.03	0.10	0.22	0.02	0.09
11	0.04	0.17	0.33	0.48	0.03	0.11	0.22	0.02	0.09
12.375	0.04	0.18	0.33	0.49	0.03	0.11	0.23	0.03	0.10
13.75	0.04	0.18	0.34	0.50	0.03	0.12	0.24	0.03	0.10
15.125	0.04	0.18	0.34	0.51	0.03	0.12	0.24	0.03	0.10
16.5	0.04	0.18	0.35	0.51	0.03	0.12	0.25	0.03	0.10
17.875	0.04	0.19	0.35	0.52	0.03	0.12	0.25	0.03	0.11
19.25	0.04	0.19	0.35	0.52	0.03	0.12	0.25	0.03	0.11
20.625	0.04	0.19	0.35	0.53	0.03	0.12	0.26	0.03	0.11
22	0.04	0.19	0.36	0.53	0.03	0.12	0.26	0.03	0.11
23.375	0.04	0.19	0.36	0.53	0.03	0.13	0.26	0.03	0.11
24.75	0.04	0.19	0.36	0.53	0.03	0.13	0.26	0.03	0.11
26.125	0.04	0.19	0.36	0.54	0.03	0.13	0.26	0.03	0.11
27.5	0.04	0.19	0.36	0.54	0.03	0.13	0.26	0.03	0.11
28.875	0.04	0.19	0.36	0.54	0.03	0.13	0.27	0.03	0.11
30.25	0.04	0.19	0.37	0.54	0.03	0.13	0.27	0.03	0.11
31.625	0.04	0.19	0.37	0.54	0.03	0.13	0.27	0.03	0.11
33	0.04	0.20	0.37	0.54	0.03	0.13	0.27	0.03	0.12
34.375	0.04	0.20	0.37	0.55	0.03	0.13	0.27	0.03	0.12
35.75	0.04	0.20	0.37	0.55	0.03	0.13	0.27	0.03	0.12
37.125		0.20	0.37	0.55	0.03	0.13	0.27	0.03	0.12
38.5		0.20	0.37	0.55	0.03	0.13	0.27	0.03	0.12
39.875		0.20	0.37	0.55	0.03	0.13	0.27	0.03	0.12
41.25		0.20	0.37	0.55	0.03	0.13	0.27	0.03	0.12
42.625		0.20	0.37	0.55	0.03	0.13	0.27	0.03	0.12
44		0.20	0.37	0.55	0.03	0.13	0.27	0.03	0.12
45.375		0.20	0.37	0.55	0.03	0.13	0.27	0.03	0.12
46.75		0.20	0.37	0.55	0.03	0.13	0.28	0.03	0.12
48.125		0.20	0.37	0.55	0.03	0.13	0.28	0.03	0.12
49.5			0.37	0.56		0.13	0.28	0.03	0.12
50.875			0.38	0.56		0.13	0.28	0.03	0.12
52.25			0.38	0.56		0.13	0.28	0.03	0.12
53.625			0.38	0.56		0.13	0.28	0.03	0.12
55			0.38	0.56		0.13	0.28	0.03	0.12
56.375			0.38	0.56		0.13	0.28	0.03	0.12
57.75			0.38	0.56		0.13	0.28	0.03	0.12
59.125			0.38	0.56		0.14	0.28	0.03	0.12
60.5			0.38	0.56		0.14	0.28	0.03	0.12
61.875			0.38	0.56		0.14	0.28	0.03	0.12
63.25			0.38	0.56		0.14	0.28	0.03	0.12
64.625				0.56			0.28		0.12
66				0.56			0.28		0.12
67.375				0.56			0.28		0.12
68.75				0.56			0.28		0.12
70.125				0.56			0.28		0.12
71.5				0.56			0.28		0.12
72.875				0.56			0.28		0.12
74.25				0.56			0.28		0.12
75.625				0.56			0.28		0.12
77				0.56			0.28		0.12

Table M16.2-3C Solid Sawn Timbers

Rating	1-HOUR				1.5-HOUR			2-HOUR	
Beam Width	5.5	7.5	9.5	11.5	7.5	9.5	11.5	9.5	11.5
Beam Depth	Design Load Ratio, R_s								
5.5	0.06								
7.5	0.06	0.22			0.05				
9.5	0.07	0.23	0.39		0.05	0.16		0.05	
11.5	0.07	0.24	0.41	0.56	0.06	0.17	0.29	0.05	0.13
13.5	0.07	0.25	0.42	0.57	0.06	0.18	0.30	0.06	0.14
15.5	0.07	0.25	0.43	0.58	0.06	0.18	0.31	0.06	0.15
17.5	0.08	0.26	0.44	0.59	0.06	0.18	0.31	0.06	0.15
19.5	0.08	0.26	0.44	0.60	0.07	0.19	0.32	0.06	0.16
21.5	0.08	0.26	0.45	0.60	0.07	0.19	0.32	0.06	0.16
23.5	0.08	0.26	0.45	0.61	0.07	0.19	0.33	0.07	0.16

Table M16.2-3B Western Species Structural Glued Laminated Timbers

Rating	1-HOUR				1.5-HOUR			2-HOUR	
Beam Width	5.125	6.75	8.75	10.75	6.75	8.75	10.75	8.75	10.75
Beam Depth	Design Load Ratio, R_s								
6	0.04								
7.5	0.04	0.16			0.02				
9	0.04	0.17	0.33		0.03	0.10		0.03	
10.5	0.04	0.17	0.34	0.49	0.03	0.11	0.22	0.03	0.10
12	0.05	0.18	0.35	0.51	0.03	0.11	0.23	0.03	0.10
13.5	0.05	0.18	0.36	0.52	0.03	0.11	0.24	0.03	0.11
15	0.05	0.18	0.36	0.53	0.03	0.12	0.24	0.03	0.11
16.5	0.05	0.18	0.37	0.53	0.03	0.12	0.25	0.03	0.11
18	0.05	0.19	0.37	0.54	0.03	0.12	0.25	0.04	0.12
19.5	0.05	0.19	0.38	0.54	0.03	0.12	0.25	0.04	0.12
21	0.05	0.19	0.38	0.55	0.03	0.12	0.26	0.04	0.12
22.5	0.05	0.19	0.38	0.55	0.03	0.13	0.26	0.04	0.12
24	0.05	0.19	0.38	0.55	0.03	0.13	0.26	0.04	0.12
25.5	0.05	0.19	0.38	0.56	0.03	0.13	0.26	0.04	0.12
27	0.05	0.19	0.39	0.56	0.03	0.13	0.26	0.04	0.13
28.5	0.05	0.19	0.39	0.56	0.03	0.13	0.27	0.04	0.13
30	0.05	0.19	0.39	0.56	0.03	0.13	0.27	0.04	0.13
31.5	0.05	0.19	0.39	0.56	0.03	0.13	0.27	0.04	0.13
33	0.05	0.20	0.39	0.56	0.03	0.13	0.27	0.04	0.13
34.5	0.05	0.20	0.39	0.57	0.03	0.13	0.27	0.04	0.13
36	0.05	0.20	0.39	0.57	0.03	0.13	0.27	0.04	0.13
37.5		0.20	0.39	0.57	0.03	0.13	0.27	0.04	0.13
39		0.20	0.39	0.57	0.03	0.13	0.27	0.04	0.13
40.5		0.20	0.40	0.57	0.03	0.13	0.27	0.04	0.13
42		0.20	0.40	0.57	0.03	0.13	0.27	0.04	0.13
43.5		0.20	0.40	0.57	0.03	0.13	0.27	0.04	0.13
45		0.20	0.40	0.57	0.03	0.13	0.27	0.04	0.13
46.5		0.20	0.40	0.57	0.03	0.13	0.28	0.04	0.13
48		0.20	0.40	0.57	0.03	0.13	0.28	0.04	0.13
49.5			0.40	0.58	0.03	0.13	0.28	0.04	0.13
51			0.40	0.58	0.03	0.13	0.28	0.04	0.13
52.5			0.40	0.58	0.03	0.13	0.28	0.04	0.13
54			0.40	0.58		0.13	0.28	0.04	0.13
55.5			0.40	0.58		0.13	0.28	0.04	0.13
57			0.40	0.58		0.13	0.28	0.04	0.13
58.5			0.40	0.58		0.13	0.28	0.04	0.13
60			0.40	0.58		0.14	0.28	0.04	0.13
61.5			0.40	0.58		0.14	0.28	0.04	0.13
63			0.40	0.58		0.14	0.28	0.04	0.13
64.5				0.58		0.14	0.28		0.13
66				0.58		0.14	0.28		0.13
67.5				0.58		0.14	0.28		0.13
69				0.58		0.14	0.28		0.14
70.5				0.58			0.28		0.14
72				0.58			0.28		0.14
73.5				0.58			0.28		0.14
75				0.58			0.28		0.14
76.5				0.58			0.28		0.14
78				0.58			0.28		0.14
79.5				0.58			0.28		0.14
81				0.58			0.28		0.14

Notes:
1. Tabulated values assume bending in the width direction.
2. Tabulated values conservatively assume column buckling failure. For relatively short, highly loaded columns, a more rigorous analysis using the *NDS* provisions will increase the design load ratio, R_s.

Table M16.2-4 Design Load Ratios for Compression Members Exposed on Three Sides

(Structural Calculations at Standard Reference Conditions: $C_M = 1.0$, $C_t = 1.0$, $C_i = 1.0$)

(Protected Surface in Width Direction)

Table M16.2-4A Southern Pine Structural Glued Laminated Timbers

Rating	1-HOUR				1.5-HOUR			2-HOUR	
Beam Width	5	6.75	8.5	10.5	6.75	8.5	10.5	8.5	10.5
Beam Depth	Design Load Ratio, R_s								
5.5	0.18								
6.875	0.25	0.38			0.14				
8.25	0.30	0.45	0.56		0.20	0.28		0.12	
9.625	0.33	0.50	0.62	0.72	0.24	0.34	0.43	0.17	0.24
11	0.36	0.54	0.67	0.78	0.28	0.39	0.49	0.21	0.29
12.375	0.38	0.57	0.70	0.82	0.30	0.42	0.53	0.25	0.34
13.75	0.39	0.59	0.73	0.85	0.32	0.45	0.57	0.27	0.37
15.125	0.41	0.61	0.76	0.88	0.34	0.48	0.60	0.29	0.40
16.5	0.42	0.63	0.78	0.90	0.35	0.50	0.62	0.31	0.43
17.875	0.42	0.64	0.79	0.92	0.36	0.51	0.65	0.32	0.45
19.25	0.43	0.65	0.81	0.94	0.37	0.53	0.66	0.34	0.47
20.625	0.44	0.66	0.82	0.95	0.38	0.54	0.68	0.35	0.48
22	0.45	0.67	0.83	0.97	0.39	0.55	0.69	0.36	0.49
23.375	0.45	0.68	0.84	0.98	0.40	0.56	0.70	0.37	0.51
24.75	0.45	0.68	0.85	0.99	0.40	0.57	0.72	0.37	0.52
26.125	0.46	0.69	0.86	1.00	0.41	0.58	0.73	0.38	0.53
27.5	0.46	0.70	0.86	1.00	0.41	0.58	0.73	0.39	0.53
28.875	0.47	0.70	0.87	1.00	0.42	0.59	0.74	0.39	0.54
30.25	0.47	0.71	0.88	1.00	0.42	0.59	0.75	0.40	0.55
31.625	0.47	0.71	0.88	1.00	0.43	0.60	0.75	0.40	0.55
33	0.47	0.71	0.89	1.00	0.43	0.60	0.76	0.41	0.56
34.375	0.48	0.72	0.89	1.00	0.43	0.61	0.77	0.41	0.57
35.75	0.48	0.72	0.89	1.00	0.43	0.61	0.77	0.41	0.57
37.125		0.72	0.90	1.00	0.44	0.62	0.78	0.42	0.57
38.5		0.73	0.90	1.00	0.44	0.62	0.78	0.42	0.58
39.875		0.73	0.90	1.00	0.44	0.62	0.78	0.42	0.58
41.25		0.73	0.91	1.00	0.44	0.63	0.79	0.43	0.59
42.625		0.73	0.91	1.00	0.45	0.63	0.79	0.43	0.59
44		0.74	0.91	1.00	0.45	0.63	0.79	0.43	0.59
45.375		0.74	0.92	1.00	0.45	0.63	0.80	0.43	0.60
46.75		0.74	0.92	1.00	0.45	0.64	0.80	0.43	0.60
48.125		0.74	0.92	1.00	0.45	0.64	0.80	0.44	0.60
49.5			0.92	1.00		0.64	0.81	0.44	0.60
50.875			0.92	1.00		0.64	0.81	0.44	0.61
52.25			0.93	1.00		0.64	0.81	0.44	0.61
53.625			0.93	1.00		0.65	0.81	0.44	0.61
55			0.93	1.00		0.65	0.82	0.44	0.61
56.375			0.93	1.00		0.65	0.82	0.45	0.62
57.75			0.93	1.00		0.65	0.82	0.45	0.62
59.125			0.93	1.00		0.65	0.82	0.45	0.62
60.5			0.94	1.00		0.65	0.82	0.45	0.62
61.875			0.94	1.00		0.66	0.82	0.45	0.62
63.25			0.94	1.00		0.66	0.83	0.45	0.62
64.625				1.00			0.83		0.62
66				1.00			0.83		0.63
67.375				1.00			0.83		0.63
68.75				1.00			0.83		0.63
70.125				1.00			0.83		0.63
71.5				1.00			0.83		0.63
72.875				1.00			0.84		0.63
74.25				1.00			0.84		0.63
75.625				1.00			0.84		0.63
77				1.00			0.84		0.64

Table M16.2-4B Western Species Structural Glued Laminated Timbers

Rating	1-HOUR				1.5-HOUR			2-HOUR	
Beam Width	5.125	6.75	8.75	10.75	6.75	8.75	10.75	8.75	10.75
Beam Depth	Design Load Ratio, R_s								
6	0.22								
7.5	0.29	0.42			0.17				
9	0.33	0.48	0.61		0.22	0.32		0.16	
10.5	0.36	0.53	0.67	0.77	0.26	0.37	0.47	0.21	0.28
12	0.39	0.56	0.71	0.82	0.29	0.42	0.52	0.25	0.34
13.5	0.41	0.59	0.75	0.86	0.32	0.45	0.56	0.28	0.38
15	0.42	0.61	0.77	0.89	0.34	0.48	0.60	0.31	0.41
16.5	0.43	0.63	0.80	0.92	0.35	0.50	0.62	0.33	0.44
18	0.44	0.64	0.81	0.94	0.37	0.51	0.65	0.34	0.48
19.5	0.45	0.65	0.83	0.96	0.38	0.53	0.67	0.36	0.48
21	0.46	0.66	0.84	0.97	0.39	0.54	0.68	0.37	0.50
22.5	0.47	0.67	0.85	0.98	0.39	0.55	0.70	0.38	0.51
24	0.47	0.68	0.86	1.00	0.40	0.56	0.71	0.39	0.53
25.5	0.48	0.69	0.87	1.00	0.41	0.57	0.72	0.40	0.54
27	0.48	0.69	0.88	1.00	0.41	0.58	0.73	0.40	0.55
28.5	0.48	0.70	0.89	1.00	0.42	0.59	0.74	0.41	0.56
30	0.49	0.70	0.90	1.00	0.42	0.59	0.75	0.42	0.56
31.5	0.49	0.71	0.90	1.00	0.43	0.60	0.75	0.42	0.57
33	0.49	0.71	0.91	1.00	0.43	0.60	0.76	0.43	0.58
34.5	0.50	0.72	0.91	1.00	0.43	0.61	0.77	0.43	0.58
36	0.50	0.72	0.92	1.00	0.44	0.61	0.77	0.44	0.59
37.5		0.72	0.92	1.00	0.44	0.62	0.78	0.44	0.59
39		0.73	0.92	1.00	0.44	0.62	0.78	0.44	0.60
40.5		0.73	0.93	1.00	0.44	0.62	0.79	0.45	0.60
42		0.73	0.93	1.00	0.45	0.63	0.79	0.45	0.61
43.5		0.73	0.93	1.00	0.45	0.63	0.79	0.45	0.61
45		0.74	0.94	1.00	0.45	0.63	0.80	0.45	0.61
46.5		0.74	0.94	1.00	0.45	0.64	0.80	0.46	0.62
48		0.74	0.94	1.00	0.45	0.64	0.80	0.46	0.62
49.5			0.94	1.00		0.64	0.81	0.46	0.62
51			0.95	1.00		0.64	0.81	0.46	0.63
52.5			0.95	1.00		0.64	0.81	0.46	0.63
54			0.95	1.00		0.65	0.81	0.47	0.63
55.5			0.95	1.00		0.65	0.82	0.47	0.63
57			0.95	1.00		0.65	0.82	0.47	0.63
58.5			0.95	1.00		0.65	0.82	0.47	0.64
60			0.96	1.00		0.65	0.82	0.47	0.64
61.5			0.96	1.00		0.65	0.82	0.47	0.64
63			0.96	1.00		0.66	0.83	0.48	0.64
64.5				1.00		0.66	0.83		0.64
66				1.00		0.66	0.83		0.65
67.5				1.00		0.66	0.83		0.65
69				1.00		0.66	0.83		0.65
70.5				1.00			0.83		0.65
72				1.00			0.83		0.65
73.5				1.00			0.84		0.65
75				1.00			0.84		0.65
76.5				1.00			0.84		0.65
78				1.00			0.84		0.66
79.5				1.00			0.84		0.66
81				1.00			0.84		0.66

Table M16.2-4C Solid Sawn Timbers

Rating	1-HOUR				1.5-HOUR			2-HOUR	
Beam Width	5.5	7.5	9.5	11.5	7.5	9.5	11.5	9.5	11.5
Beam Depth	Design Load Ratio, R_s								
5.5	0.21								
7.5	0.32	0.46			0.20				
9.5	0.38	0.55	0.67		0.28	0.38		0.20	
11.5	0.42	0.61	0.74	0.84	0.34	0.46	0.55	0.27	0.35
13.5	0.45	0.65	0.79	0.89	0.38	0.51	0.61	0.32	0.41
15.5	0.47	0.68	0.83	0.94	0.41	0.55	0.66	0.36	0.46
17.5	0.49	0.71	0.86	0.97	0.43	0.58	0.69	0.38	0.49
19.5	0.50	0.73	0.88	0.99	0.45	0.60	0.72	0.41	0.52
21.5	0.51	0.74	0.90	1.00	0.46	0.62	0.75	0.43	0.55
23.5	0.52	0.75	0.92	1.00	0.47	0.64	0.77	0.44	0.57

Notes:
1. Tabulated values assume bending in the width direction.
2. Tabulated values conservatively assume column buckling failure. For relatively short, highly loaded columns, a more rigorous analysis using the *NDS* provisions will increase the design load ratio, R_s.

Table M16.2-5 Design Load Ratios for Compression Members Exposed on Four Sides

(Structural Calculations at Standard Reference Conditions: $C_M = 1.0$, $C_t = 1.0$, $C_i = 1.0$)

Table M16.2-5A Southern Pine Structural Glued Laminated Timbers

Rating	1-HOUR				1.5-HOUR			2-HOUR	
Beam Width	5	6.75	8.5	10.5	6.75	8.5	10.5	8.5	10.5
Beam Depth	Design Load Ratio, R_s								
5.5	0.02								
6.875	0.02	0.10			0.01				
8.25	0.03	0.12	0.22		0.01	0.06		0.01	
9.625	0.03	0.13	0.24	0.36	0.02	0.07	0.14	0.01	0.04
11	0.03	0.14	0.26	0.39	0.02	0.08	0.16	0.01	0.05
12.375	0.03	0.15	0.28	0.41	0.02	0.08	0.17	0.02	0.06
13.75	0.03	0.15	0.29	0.43	0.02	0.09	0.18	0.02	0.07
15.125	0.03	0.16	0.30	0.44	0.02	0.09	0.19	0.02	0.07
16.5	0.03	0.16	0.30	0.45	0.02	0.10	0.20	0.02	0.08
17.875	0.04	0.16	0.31	0.46	0.03	0.10	0.21	0.02	0.08
19.25	0.04	0.17	0.32	0.47	0.03	0.10	0.22	0.02	0.09
20.625	0.04	0.17	0.32	0.48	0.03	0.11	0.22	0.02	0.09
22	0.04	0.17	0.33	0.48	0.03	0.11	0.22	0.02	0.09
23.375	0.04	0.17	0.33	0.49	0.03	0.11	0.23	0.02	0.09
24.75	0.04	0.18	0.33	0.49	0.03	0.11	0.23	0.03	0.10
26.125	0.04	0.18	0.34	0.50	0.03	0.11	0.24	0.03	0.10
27.5	0.04	0.18	0.34	0.50	0.03	0.12	0.24	0.03	0.10
28.875	0.04	0.18	0.34	0.50	0.03	0.12	0.24	0.03	0.10
30.25	0.04	0.18	0.34	0.51	0.03	0.12	0.24	0.03	0.10
31.625	0.04	0.18	0.34	0.51	0.03	0.12	0.24	0.03	0.10
33	0.04	0.18	0.35	0.51	0.03	0.12	0.25	0.03	0.10
34.375	0.04	0.18	0.35	0.52	0.03	0.12	0.25	0.03	0.10
35.75	0.04	0.19	0.35	0.52	0.03	0.12	0.25	0.03	0.11
37.125		0.19	0.35	0.52	0.03	0.12	0.25	0.03	0.11
38.5		0.19	0.35	0.52	0.03	0.12	0.25	0.03	0.11
39.875		0.19	0.35	0.52	0.03	0.12	0.25	0.03	0.11
41.25		0.19	0.35	0.53	0.03	0.12	0.26	0.03	0.11
42.625		0.19	0.36	0.53	0.03	0.12	0.26	0.03	0.11
44		0.19	0.36	0.53	0.03	0.12	0.26	0.03	0.11
45.375		0.19	0.36	0.53	0.03	0.13	0.26	0.03	0.11
46.75		0.19	0.36	0.53	0.03	0.13	0.26	0.03	0.11
48.125		0.19	0.36	0.53	0.03	0.13	0.26	0.03	0.11
49.5			0.36	0.53		0.13	0.26	0.03	0.11
50.875			0.36	0.54		0.13	0.26	0.03	0.11
52.25			0.36	0.54		0.13	0.26	0.03	0.11
53.625			0.36	0.54		0.13	0.26	0.03	0.11
55			0.36	0.54		0.13	0.26	0.03	0.11
56.375			0.36	0.54		0.13	0.27	0.03	0.11
57.75			0.36	0.54		0.13	0.27	0.03	0.11
59.125			0.37	0.54		0.13	0.27	0.03	0.11
60.5			0.37	0.54		0.13	0.27	0.03	0.11
61.875			0.37	0.54		0.13	0.27	0.03	0.11
63.25			0.37	0.54		0.13	0.27	0.03	0.11
64.625				0.54			0.27		0.12
66				0.54			0.27		0.12
67.375				0.55			0.27		0.12
68.75				0.55			0.27		0.12
70.125				0.55			0.27		0.12
71.5				0.55			0.27		0.12
72.875				0.55			0.27		0.12
74.25				0.55			0.27		0.12
75.625				0.55			0.27		0.12
77				0.55			0.27		0.12

Table M16.2-5B Western Species Structural Glued Laminated Timbers

Rating	1-HOUR				1.5-HOUR			2-HOUR	
Beam Width	5.125	6.75	8.75	10.75	6.75	8.75	10.75	8.75	10.75
Beam Depth	Design Load Ratio, R_s								
6	0.02								
7.5	0.03	0.11			0.01				
9	0.03	0.12	0.25		0.02	0.06		0.01	
10.5	0.04	0.14	0.27	0.39	0.02	0.07	0.15	0.02	0.06
12	0.04	0.14	0.29	0.42	0.02	0.08	0.17	0.02	0.07
13.5	0.04	0.15	0.30	0.44	0.02	0.09	0.18	0.02	0.08
15	0.04	0.16	0.31	0.45	0.02	0.09	0.19	0.03	0.08
16.5	0.04	0.16	0.32	0.47	0.02	0.10	0.20	0.03	0.09
18	0.04	0.17	0.33	0.48	0.03	0.10	0.21	0.03	0.09
19.5	0.04	0.17	0.34	0.49	0.03	0.10	0.22	0.03	0.10
21	0.04	0.17	0.34	0.49	0.03	0.11	0.22	0.03	0.10
22.5	0.04	0.17	0.35	0.50	0.03	0.11	0.23	0.03	0.10
24	0.05	0.18	0.35	0.51	0.03	0.11	0.23	0.03	0.10
25.5	0.05	0.18	0.36	0.51	0.03	0.11	0.23	0.03	0.11
27	0.05	0.18	0.36	0.52	0.03	0.11	0.24	0.03	0.11
28.5	0.05	0.18	0.36	0.52	0.03	0.12	0.24	0.03	0.11
30	0.05	0.18	0.36	0.53	0.03	0.12	0.24	0.03	0.11
31.5	0.05	0.18	0.37	0.53	0.03	0.12	0.24	0.03	0.11
33	0.05	0.18	0.37	0.53	0.03	0.12	0.25	0.03	0.11
34.5	0.05	0.18	0.37	0.53	0.03	0.12	0.25	0.04	0.12
36	0.05	0.19	0.37	0.54	0.03	0.12	0.25	0.04	0.12
37.5		0.19	0.37	0.54	0.03	0.12	0.25	0.04	0.12
39		0.19	0.38	0.54	0.03	0.12	0.25	0.04	0.12
40.5		0.19	0.38	0.54	0.03	0.12	0.25	0.04	0.12
42		0.19	0.38	0.55	0.03	0.12	0.26	0.04	0.12
43.5		0.19	0.38	0.55	0.03	0.12	0.26	0.04	0.12
45		0.19	0.38	0.55	0.03	0.13	0.26	0.04	0.12
46.5		0.19	0.38	0.55	0.03	0.13	0.26	0.04	0.12
48		0.19	0.38	0.55	0.03	0.13	0.26	0.04	0.12
49.5			0.38	0.55	0.03	0.13	0.26	0.04	0.12
51			0.38	0.56	0.03	0.13	0.26	0.04	0.12
52.5			0.39	0.56	0.03	0.13	0.26	0.04	0.12
54			0.39	0.56		0.13	0.26	0.04	0.13
55.5			0.39	0.56		0.13	0.26	0.04	0.13
57			0.39	0.56		0.13	0.27	0.04	0.13
58.5			0.39	0.56		0.13	0.27	0.04	0.13
60			0.39	0.56		0.13	0.27	0.04	0.13
61.5			0.39	0.56		0.13	0.27	0.04	0.13
63			0.39	0.56		0.13	0.27	0.04	0.13
64.5				0.56		0.13	0.27		0.13
66				0.56		0.13	0.27		0.13
67.5				0.57		0.13	0.27		0.13
69				0.57		0.13	0.27		0.13
70.5				0.57			0.27		0.13
72				0.57			0.27		0.13
73.5				0.57			0.27		0.13
75				0.57			0.27		0.13
76.5				0.57			0.27		0.13
78				0.57			0.27		0.13
79.5				0.57			0.27		0.13
81				0.57			0.27		0.13

Notes:
1. Tabulated values assume bending in the width direction.
2. Tabulated values conservatively assume column buckling failure. For relatively short, highly loaded columns, a more rigorous analysis using the *NDS* provisions will increase the design load ratio, R_s.

Table M16.2-5C Solid Sawn Timbers

Rating	1-HOUR				1.5-HOUR			2-HOUR	
Beam Width	5.5	7.5	9.5	11.5	7.5	9.5	11.5	9.5	11.5
Beam Depth	Design Load Ratio, R_s								
5.5	0.03								
7.5	0.04	0.15			0.02				
9.5	0.05	0.18	0.30		0.04	0.10		0.03	
11.5	0.06	0.20	0.33	0.45	0.04	0.12	0.21	0.03	0.08
13.5	0.06	0.21	0.36	0.48	0.05	0.14	0.23	0.04	0.10
15.5	0.06	0.22	0.37	0.51	0.05	0.15	0.25	0.04	0.11
17.5	0.07	0.23	0.39	0.52	0.05	0.15	0.26	0.05	0.12
19.5	0.07	0.23	0.40	0.54	0.06	0.16	0.27	0.05	0.13
21.5	0.07	0.24	0.40	0.55	0.06	0.16	0.28	0.05	0.13
23.5	0.07	0.24	0.41	0.56	0.06	0.17	0.29	0.06	0.14

Table M16.2-6 Design Load Ratios for Tension Members Exposed on Three Sides

(Structural Calculations at Standard Reference Conditions: $C_D = 1.0$, $C_M = 1.0$, $C_t = 1.0$, $C_i = 1.0$)

(Protected Surface in Depth Direction)

Table M16.2-6A Southern Pine Structural Glued Laminated Timbers

Rating	1-HOUR				1.5-HOUR			2-HOUR	
Beam Width	5	6.75	8.5	10.5	6.75	8.5	10.5	8.5	10.5
Beam Depth	Design Load Ratio, R_s								
5.5	0.54	0.89	1.00	1.00	0.40	0.64	0.81	0.31	0.48
6.875	0.59	0.98	1.00	1.00	0.47	0.75	0.95	0.39	0.61
8.25	0.62	1.00	1.00	1.00	0.51	0.82	1.00	0.45	0.70
9.625	0.65	1.00	1.00	1.00	0.54	0.87	1.00	0.49	0.76
11	0.67	1.00	1.00	1.00	0.57	0.91	1.00	0.52	0.81
12.375	0.68	1.00	1.00	1.00	0.59	0.93	1.00	0.54	0.84
13.75	0.69	1.00	1.00	1.00	0.60	0.96	1.00	0.56	0.87
15.125	0.70	1.00	1.00	1.00	0.61	0.98	1.00	0.58	0.90
16.5	0.71	1.00	1.00	1.00	0.62	0.99	1.00	0.59	0.92
17.875	0.72	1.00	1.00	1.00	0.63	1.00	1.00	0.60	0.93
19.25	0.72	1.00	1.00	1.00	0.64	1.00	1.00	0.61	0.95
20.625	0.73	1.00	1.00	1.00	0.65	1.00	1.00	0.62	0.96
22	0.73	1.00	1.00	1.00	0.65	1.00	1.00	0.62	0.97
23.375	0.74	1.00	1.00	1.00	0.66	1.00	1.00	0.63	0.98
24.75	0.74	1.00	1.00	1.00	0.66	1.00	1.00	0.64	0.99
26.125	0.74	1.00	1.00	1.00	0.67	1.00	1.00	0.64	1.00
27.5	0.75	1.00	1.00	1.00	0.67	1.00	1.00	0.65	1.00
28.875	0.75	1.00	1.00	1.00	0.67	1.00	1.00	0.65	1.00
30.25	0.75	1.00	1.00	1.00	0.68	1.00	1.00	0.65	1.00
31.625	0.75	1.00	1.00	1.00	0.68	1.00	1.00	0.66	1.00
33	0.75	1.00	1.00	1.00	0.68	1.00	1.00	0.66	1.00
34.375	0.76	1.00	1.00	1.00	0.68	1.00	1.00	0.66	1.00
35.75	0.76	1.00	1.00	1.00	0.68	1.00	1.00	0.66	1.00
37.125		1.00	1.00	1.00	0.69	1.00	1.00	0.67	1.00
38.5		1.00	1.00	1.00	0.69	1.00	1.00	0.67	1.00
39.875		1.00	1.00	1.00	0.69	1.00	1.00	0.67	1.00
41.25		1.00	1.00	1.00	0.69	1.00	1.00	0.67	1.00
42.625		1.00	1.00	1.00	0.69	1.00	1.00	0.68	1.00
44		1.00	1.00	1.00	0.69	1.00	1.00	0.68	1.00
45.375		1.00	1.00	1.00	0.70	1.00	1.00	0.68	1.00
46.75		1.00	1.00	1.00	0.70	1.00	1.00	0.68	1.00
48.125		1.00	1.00	1.00	0.70	1.00	1.00	0.68	1.00
49.5			1.00	1.00		1.00	1.00	0.68	1.00
50.875			1.00	1.00		1.00	1.00	0.68	1.00
52.25			1.00	1.00		1.00	1.00	0.69	1.00
53.625			1.00	1.00		1.00	1.00	0.69	1.00
55			1.00	1.00		1.00	1.00	0.69	1.00
56.375			1.00	1.00		1.00	1.00	0.69	1.00
57.75			1.00	1.00		1.00	1.00	0.69	1.00
59.125			1.00	1.00		1.00	1.00	0.69	1.00
60.5			1.00	1.00		1.00	1.00	0.69	1.00
61.875			1.00	1.00		1.00	1.00	0.69	1.00
63.25			1.00	1.00		1.00	1.00	0.69	1.00
64.625				1.00			1.00		1.00
66				1.00			1.00		1.00
67.375				1.00			1.00		1.00
68.75				1.00			1.00		1.00
70.125				1.00			1.00		1.00
71.5				1.00			1.00		1.00
72.875				1.00			1.00		1.00
74.25				1.00			1.00		1.00
75.625				1.00			1.00		1.00
77				1.00			1.00		1.00

Table M16.2-6B Western Species Structural Glued Laminated Timbers

Rating	1-HOUR				1.5-HOUR			2-HOUR	
Beam Width	5.125	6.75	8.75	10.75	6.75	8.75	10.75	8.75	10.75
Beam Depth	Design Load Ratio, R_s								
6	0.59	0.93	1.00	1.00	0.43	0.71	0.89	0.37	0.55
7.5	0.64	1.00	1.00	1.00	0.49	0.81	1.00	0.46	0.68
9	0.68	1.00	1.00	1.00	0.53	0.88	1.00	0.51	0.76
10.5	0.70	1.00	1.00	1.00	0.56	0.93	1.00	0.55	0.82
12	0.72	1.00	1.00	1.00	0.58	0.97	1.00	0.58	0.86
13.5	0.73	1.00	1.00	1.00	0.60	0.99	1.00	0.60	0.90
15	0.75	1.00	1.00	1.00	0.61	1.00	1.00	0.62	0.93
16.5	0.76	1.00	1.00	1.00	0.62	1.00	1.00	0.64	0.95
18	0.76	1.00	1.00	1.00	0.63	1.00	1.00	0.65	0.97
19.5	0.77	1.00	1.00	1.00	0.64	1.00	1.00	0.66	0.98
21	0.78	1.00	1.00	1.00	0.65	1.00	1.00	0.67	1.00
22.5	0.78	1.00	1.00	1.00	0.65	1.00	1.00	0.68	1.00
24	0.78	1.00	1.00	1.00	0.66	1.00	1.00	0.69	1.00
25.5	0.79	1.00	1.00	1.00	0.66	1.00	1.00	0.69	1.00
27	0.79	1.00	1.00	1.00	0.67	1.00	1.00	0.70	1.00
28.5	0.79	1.00	1.00	1.00	0.67	1.00	1.00	0.70	1.00
30	0.80	1.00	1.00	1.00	0.68	1.00	1.00	0.71	1.00
31.5	0.80	1.00	1.00	1.00	0.68	1.00	1.00	0.71	1.00
33	0.80	1.00	1.00	1.00	0.68	1.00	1.00	0.71	1.00
34.5	0.80	1.00	1.00	1.00	0.68	1.00	1.00	0.72	1.00
36	0.81	1.00	1.00	1.00	0.69	1.00	1.00	0.72	1.00
37.5		1.00	1.00	1.00	0.69	1.00	1.00	0.72	1.00
39		1.00	1.00	1.00	0.69	1.00	1.00	0.73	1.00
40.5		1.00	1.00	1.00	0.69	1.00	1.00	0.73	1.00
42		1.00	1.00	1.00	0.69	1.00	1.00	0.73	1.00
43.5		1.00	1.00	1.00	0.69	1.00	1.00	0.73	1.00
45		1.00	1.00	1.00	0.70	1.00	1.00	0.73	1.00
46.5		1.00	1.00	1.00	0.70	1.00	1.00	0.74	1.00
48		1.00	1.00	1.00	0.70	1.00	1.00	0.74	1.00
49.5			1.00	1.00		1.00	1.00	0.74	1.00
51			1.00	1.00		1.00	1.00	0.74	1.00
52.5			1.00	1.00		1.00	1.00	0.74	1.00
54			1.00	1.00		1.00	1.00	0.74	1.00
55.5			1.00	1.00		1.00	1.00	0.74	1.00
57			1.00	1.00		1.00	1.00	0.75	1.00
58.5			1.00	1.00		1.00	1.00	0.75	1.00
60			1.00	1.00		1.00	1.00	0.75	1.00
61.5			1.00	1.00		1.00	1.00	0.75	1.00
63			1.00	1.00		1.00	1.00	0.75	1.00
64.5				1.00			1.00		1.00
66				1.00			1.00		1.00
67.5				1.00			1.00		1.00
69				1.00			1.00		1.00
70.5				1.00			1.00		1.00
72				1.00			1.00		1.00
73.5				1.00			1.00		1.00
75				1.00			1.00		1.00
76.5				1.00			1.00		1.00
78				1.00			1.00		1.00
79.5				1.00			1.00		1.00
81				1.00			1.00		1.00

Table M16.2-6C Solid Sawn Timbers

Rating	1-HOUR				1.5-HOUR			2-HOUR	
Beam Width	5.5	7.5	9.5	11.5	7.5	9.5	11.5	9.5	11.5
Beam Depth	Design Load Ratio, R_s								
5.5	0.66	1.00	1.00	1.00	0.52	0.73	0.88	0.40	0.55
7.5	0.75	1.00	1.00	1.00	0.63	0.90	1.00	0.55	0.74
9.5	0.80	1.00	1.00	1.00	0.70	0.99	1.00	0.64	0.86
11.5	0.83	1.00	1.00	1.00	0.74	1.00	1.00	0.69	0.93
13.5	0.85	1.00	1.00	1.00	0.77	1.00	1.00	0.73	0.98
15.5	0.87	1.00	1.00	1.00	0.79	1.00	1.00	0.76	1.00
17.5	0.88	1.00	1.00	1.00	0.81	1.00	1.00	0.78	1.00
19.5	0.89	1.00	1.00	1.00	0.83	1.00	1.00	0.80	1.00
21.5	0.90	1.00	1.00	1.00	0.84	1.00	1.00	0.81	1.00
23.5	0.91	1.00	1.00	1.00	0.85	1.00	1.00	0.82	1.00

Table M16.2-7 Design Load Ratios for Tension Members Exposed on Three Sides

(Structural Calculations at Standard Reference Conditions: $C_D = 1.0$, $C_M = 1.0$, $C_t = 1.0$, $C_i = 1.0$)

(Protected Surface in Width Direction)

Table M16.2-7A Southern Pine Glued Structural Laminated Timbers

Rating	1-HOUR				1.5-HOUR			2-HOUR	
Beam Width	5	6.75	8.5	10.5	6.75	8.5	10.5	8.5	10.5
Beam Depth	Design Load Ratio, R_s								
5.5	0.63	0.72	0.78	0.82	0.16	0.18	0.20		
6.875	0.87	1.00	1.00	1.00	0.49	0.55	0.59	0.14	0.16
8.25	1.00	1.00	1.00	1.00	0.71	0.79	0.85	0.42	0.46
9.625	1.00	1.00	1.00	1.00	0.86	0.97	1.00	0.61	0.68
11	1.00	1.00	1.00	1.00	0.98	1.00	1.00	0.76	0.85
12.375	1.00	1.00	1.00	1.00	1.00	1.00	1.00	0.87	0.97
13.75	1.00	1.00	1.00	1.00	1.00	1.00	1.00	0.97	1.00
15.125	1.00	1.00	1.00	1.00	1.00	1.00	1.00	1.00	1.00
16.5	1.00	1.00	1.00	1.00	1.00	1.00	1.00	1.00	1.00
17.875	1.00	1.00	1.00	1.00	1.00	1.00	1.00	1.00	1.00
19.25	1.00	1.00	1.00	1.00	1.00	1.00	1.00	1.00	1.00
20.625	1.00	1.00	1.00	1.00	1.00	1.00	1.00	1.00	1.00
22	1.00	1.00	1.00	1.00	1.00	1.00	1.00	1.00	1.00
23.375	1.00	1.00	1.00	1.00	1.00	1.00	1.00	1.00	1.00
24.75	1.00	1.00	1.00	1.00	1.00	1.00	1.00	1.00	1.00
26.125	1.00	1.00	1.00	1.00	1.00	1.00	1.00	1.00	1.00
27.5	1.00	1.00	1.00	1.00	1.00	1.00	1.00	1.00	1.00
28.875	1.00	1.00	1.00	1.00	1.00	1.00	1.00	1.00	1.00
30.25	1.00	1.00	1.00	1.00	1.00	1.00	1.00	1.00	1.00
31.625	1.00	1.00	1.00	1.00	1.00	1.00	1.00	1.00	1.00
33	1.00	1.00	1.00	1.00	1.00	1.00	1.00	1.00	1.00
34.375	1.00	1.00	1.00	1.00	1.00	1.00	1.00	1.00	1.00
35.75	1.00	1.00	1.00	1.00	1.00	1.00	1.00	1.00	1.00
37.125		1.00	1.00	1.00	1.00	1.00	1.00	1.00	1.00
38.5		1.00	1.00	1.00	1.00	1.00	1.00	1.00	1.00
39.875		1.00	1.00	1.00	1.00	1.00	1.00	1.00	1.00
41.25		1.00	1.00	1.00	1.00	1.00	1.00	1.00	1.00
42.625		1.00	1.00	1.00	1.00	1.00	1.00	1.00	1.00
44		1.00	1.00	1.00	1.00	1.00	1.00	1.00	1.00
45.375		1.00	1.00	1.00	1.00	1.00	1.00	1.00	1.00
46.75		1.00	1.00	1.00	1.00	1.00	1.00	1.00	1.00
48.125		1.00	1.00	1.00	1.00	1.00	1.00	1.00	1.00
49.5			1.00	1.00		1.00	1.00	1.00	1.00
50.875			1.00	1.00		1.00	1.00	1.00	1.00
52.25			1.00	1.00		1.00	1.00	1.00	1.00
53.625			1.00	1.00		1.00	1.00	1.00	1.00
55			1.00	1.00		1.00	1.00	1.00	1.00
56.375			1.00	1.00		1.00	1.00	1.00	1.00
57.75			1.00	1.00		1.00	1.00	1.00	1.00
59.125			1.00	1.00		1.00	1.00	1.00	1.00
60.5			1.00	1.00		1.00	1.00	1.00	1.00
61.875			1.00	1.00		1.00	1.00	1.00	1.00
63.25			1.00	1.00		1.00	1.00	1.00	1.00
64.625				1.00			1.00		1.00
66				1.00			1.00		1.00
67.375				1.00			1.00		1.00
68.75				1.00			1.00		1.00
70.125				1.00			1.00		1.00
71.5				1.00			1.00		1.00
72.875				1.00			1.00		1.00
74.25				1.00			1.00		1.00
75.625				1.00			1.00		1.00
77				1.00			1.00		1.00

Table M16.2-7B Western Species Structural Glued Laminated Timbers

Rating	1-HOUR				1.5-HOUR			2-HOUR	
Beam Width	5.125	6.75	8.75	10.75	6.75	8.75	10.75	8.75	10.75
Beam Depth	Design Load Ratio, R_s								
6	0.74	0.84	0.91	0.95	0.30	0.34	0.36		
7.5	0.96	1.00	1.00	1.00	0.60	0.68	0.73	0.29	0.32
9	1.00	1.00	1.00	1.00	0.80	0.90	0.97	0.54	0.60
10.5	1.00	1.00	1.00	1.00	0.94	1.00	1.00	0.72	0.80
12	1.00	1.00	1.00	1.00	1.00	1.00	1.00	0.86	0.95
13.5	1.00	1.00	1.00	1.00	1.00	1.00	1.00	0.97	1.00
15	1.00	1.00	1.00	1.00	1.00	1.00	1.00	1.00	1.00
16.5	1.00	1.00	1.00	1.00	1.00	1.00	1.00	1.00	1.00
18	1.00	1.00	1.00	1.00	1.00	1.00	1.00	1.00	1.00
19.5	1.00	1.00	1.00	1.00	1.00	1.00	1.00	1.00	1.00
21	1.00	1.00	1.00	1.00	1.00	1.00	1.00	1.00	1.00
22.5	1.00	1.00	1.00	1.00	1.00	1.00	1.00	1.00	1.00
24	1.00	1.00	1.00	1.00	1.00	1.00	1.00	1.00	1.00
25.5	1.00	1.00	1.00	1.00	1.00	1.00	1.00	1.00	1.00
27	1.00	1.00	1.00	1.00	1.00	1.00	1.00	1.00	1.00
28.5	1.00	1.00	1.00	1.00	1.00	1.00	1.00	1.00	1.00
30	1.00	1.00	1.00	1.00	1.00	1.00	1.00	1.00	1.00
31.5	1.00	1.00	1.00	1.00	1.00	1.00	1.00	1.00	1.00
33	1.00	1.00	1.00	1.00	1.00	1.00	1.00	1.00	1.00
34.5	1.00	1.00	1.00	1.00	1.00	1.00	1.00	1.00	1.00
36	1.00	1.00	1.00	1.00	1.00	1.00	1.00	1.00	1.00
37.5		1.00	1.00	1.00	1.00	1.00	1.00	1.00	1.00
39		1.00	1.00	1.00	1.00	1.00	1.00	1.00	1.00
40.5		1.00	1.00	1.00	1.00	1.00	1.00	1.00	1.00
42		1.00	1.00	1.00	1.00	1.00	1.00	1.00	1.00
43.5		1.00	1.00	1.00	1.00	1.00	1.00	1.00	1.00
45		1.00	1.00	1.00	1.00	1.00	1.00	1.00	1.00
46.5		1.00	1.00	1.00	1.00	1.00	1.00	1.00	1.00
48		1.00	1.00	1.00	1.00	1.00	1.00	1.00	1.00
49.5			1.00	1.00		1.00	1.00	1.00	1.00
51			1.00	1.00		1.00	1.00	1.00	1.00
52.5			1.00	1.00		1.00	1.00	1.00	1.00
54			1.00	1.00		1.00	1.00	1.00	1.00
55.5			1.00	1.00		1.00	1.00	1.00	1.00
57			1.00	1.00		1.00	1.00	1.00	1.00
58.5			1.00	1.00		1.00	1.00	1.00	1.00
60			1.00	1.00		1.00	1.00	1.00	1.00
61.5			1.00	1.00		1.00	1.00	1.00	1.00
63			1.00	1.00		1.00	1.00	1.00	1.00
64.5				1.00			1.00		1.00
66				1.00			1.00		1.00
67.5				1.00			1.00		1.00
69				1.00			1.00		1.00
70.5				1.00			1.00		1.00
72				1.00			1.00		1.00
73.5				1.00			1.00		1.00
75				1.00			1.00		1.00
76.5				1.00			1.00		1.00
78				1.00			1.00		1.00
79.5				1.00			1.00		1.00
81				1.00			1.00		1.00

Table M16.2-7C Solid Sawn Timbers

Rating	1-HOUR				1.5-HOUR			2-HOUR	
Beam Width	5.5	7.5	9.5	11.5	7.5	9.5	11.5	9.5	11.5
Beam Depth	Design Load Ratio, R_s								
5.5	0.66	0.75	0.80	0.83	0.17	0.19	0.20		
7.5	1.00	1.00	1.00	1.00	0.63	0.70	0.74	0.30	0.32
9.5	1.00	1.00	1.00	1.00	0.90	0.99	1.00	0.64	0.69
11.5	1.00	1.00	1.00	1.00	1.00	1.00	1.00	0.86	0.93
13.5	1.00	1.00	1.00	1.00	1.00	1.00	1.00	1.00	1.00
15.5	1.00	1.00	1.00	1.00	1.00	1.00	1.00	1.00	1.00
17.5	1.00	1.00	1.00	1.00	1.00	1.00	1.00	1.00	1.00
19.5	1.00	1.00	1.00	1.00	1.00	1.00	1.00	1.00	1.00
21.5	1.00	1.00	1.00	1.00	1.00	1.00	1.00	1.00	1.00
23.5	1.00	1.00	1.00	1.00	1.00	1.00	1.00	1.00	1.00

Table M16.2-8 Design Load Ratios for Tension Members Exposed on Four Sides

(Structural Calculations at Standard Reference Conditions: $C_D = 1.0$, $C_M = 1.0$, $C_t = 1.0$, $C_i = 1.0$)

Table M16.2-8A Southern Pine Structural Glued Laminated Timbers

Rating	1-HOUR				1.5-HOUR			2-HOUR	
Beam Width	5	6.75	8.5	10.5	6.75	8.5	10.5	8.5	10.5
Beam Depth	Design Load Ratio, R_s								
5.5	0.28	0.46	0.57	0.65	0.07	0.11	0.13		
6.875	0.38	0.63	0.78	0.89	0.20	0.32	0.41	0.06	0.09
8.25	0.45	0.75	0.93	1.00	0.29	0.46	0.59	0.17	0.26
9.625	0.50	0.83	1.00	1.00	0.35	0.56	0.72	0.25	0.39
11	0.54	0.89	1.00	1.00	0.40	0.64	0.81	0.31	0.48
12.375	0.57	0.94	1.00	1.00	0.44	0.70	0.89	0.36	0.55
13.75	0.59	0.98	1.00	1.00	0.47	0.75	0.95	0.39	0.61
15.125	0.61	1.00	1.00	1.00	0.49	0.78	1.00	0.42	0.66
16.5	0.62	1.00	1.00	1.00	0.51	0.82	1.00	0.45	0.70
17.875	0.64	1.00	1.00	1.00	0.53	0.84	1.00	0.47	0.73
19.25	0.65	1.00	1.00	1.00	0.54	0.87	1.00	0.49	0.76
20.625	0.66	1.00	1.00	1.00	0.56	0.89	1.00	0.51	0.79
22	0.67	1.00	1.00	1.00	0.57	0.91	1.00	0.52	0.81
23.375	0.68	1.00	1.00	1.00	0.58	0.92	1.00	0.53	0.83
24.75	0.68	1.00	1.00	1.00	0.59	0.93	1.00	0.54	0.84
26.125	0.69	1.00	1.00	1.00	0.60	0.95	1.00	0.55	0.86
27.5	0.69	1.00	1.00	1.00	0.60	0.96	1.00	0.56	0.87
28.875	0.70	1.00	1.00	1.00	0.61	0.97	1.00	0.57	0.89
30.25	0.70	1.00	1.00	1.00	0.61	0.98	1.00	0.58	0.90
31.625	0.71	1.00	1.00	1.00	0.62	0.99	1.00	0.58	0.91
33	0.71	1.00	1.00	1.00	0.62	0.99	1.00	0.59	0.92
34.375	0.71	1.00	1.00	1.00	0.63	1.00	1.00	0.60	0.92
35.75	0.72	1.00	1.00	1.00	0.63	1.00	1.00	0.60	0.93
37.125		1.00	1.00	1.00	0.64	1.00	1.00	0.61	0.94
38.5		1.00	1.00	1.00	0.64	1.00	1.00	0.61	0.95
39.875		1.00	1.00	1.00	0.64	1.00	1.00	0.61	0.95
41.25		1.00	1.00	1.00	0.65	1.00	1.00	0.62	0.96
42.625		1.00	1.00	1.00	0.65	1.00	1.00	0.62	0.97
44		1.00	1.00	1.00	0.65	1.00	1.00	0.62	0.97
45.375		1.00	1.00	1.00	0.66	1.00	1.00	0.63	0.98
46.75		1.00	1.00	1.00	0.66	1.00	1.00	0.63	0.98
48.125		1.00	1.00	1.00	0.66	1.00	1.00	0.63	0.98
49.5			1.00	1.00		1.00	1.00	0.64	0.99
50.875			1.00	1.00		1.00	1.00	0.64	0.99
52.25			1.00	1.00		1.00	1.00	0.64	1.00
53.625			1.00	1.00		1.00	1.00	0.64	1.00
55			1.00	1.00		1.00	1.00	0.65	1.00
56.375			1.00	1.00		1.00	1.00	0.65	1.00
57.75			1.00	1.00		1.00	1.00	0.65	1.00
59.125			1.00	1.00		1.00	1.00	0.65	1.00
60.5			1.00	1.00		1.00	1.00	0.65	1.00
61.875			1.00	1.00		1.00	1.00	0.65	1.00
63.25			1.00	1.00		1.00	1.00	0.66	1.00
64.625				1.00			1.00		1.00
66				1.00			1.00		1.00
67.375				1.00			1.00		1.00
68.75				1.00			1.00		1.00
70.125				1.00			1.00		1.00
71.5				1.00			1.00		1.00
72.875				1.00			1.00		1.00
74.25				1.00			1.00		1.00
75.625				1.00			1.00		1.00
77				1.00			1.00		1.00

Table M16.2-8B Western Species Structural Glued Laminated Timbers

Rating	1-HOUR				1.5-HOUR			2-HOUR	
Beam Width	5.125	6.75	8.75	10.75	6.75	8.75	10.75	8.75	10.75
Beam Depth	Design Load Ratio, R_s								
6	0.34	0.53	0.67	0.76	0.12	0.20	0.25		
7.5	0.44	0.69	0.87	0.99	0.24	0.41	0.51	0.12	0.1
9	0.51	0.80	1.00	1.00	0.33	0.54	0.68	0.23	0.3
10.5	0.56	0.87	1.00	1.00	0.39	0.64	0.80	0.31	0.4
12	0.59	0.93	1.00	1.00	0.43	0.71	0.89	0.37	0.5
13.5	0.62	0.98	1.00	1.00	0.46	0.77	0.96	0.42	0.62
15	0.64	1.00	1.00	1.00	0.49	0.81	1.00	0.46	0.6
16.5	0.66	1.00	1.00	1.00	0.51	0.85	1.00	0.49	0.7
18	0.68	1.00	1.00	1.00	0.53	0.88	1.00	0.51	0.7
19.5	0.69	1.00	1.00	1.00	0.55	0.91	1.00	0.53	0.7
21	0.70	1.00	1.00	1.00	0.56	0.93	1.00	0.55	0.8
22.5	0.71	1.00	1.00	1.00	0.57	0.95	1.00	0.57	0.8
24	0.72	1.00	1.00	1.00	0.58	0.97	1.00	0.58	0.8
25.5	0.73	1.00	1.00	1.00	0.59	0.98	1.00	0.59	0.8
27	0.73	1.00	1.00	1.00	0.60	0.99	1.00	0.60	0.9
28.5	0.74	1.00	1.00	1.00	0.61	1.00	1.00	0.61	0.9
30	0.75	1.00	1.00	1.00	0.61	1.00	1.00	0.62	0.9
31.5	0.75	1.00	1.00	1.00	0.62	1.00	1.00	0.63	0.9
33	0.76	1.00	1.00	1.00	0.62	1.00	1.00	0.64	0.9
34.5	0.76	1.00	1.00	1.00	0.63	1.00	1.00	0.65	0.9
36	0.76	1.00	1.00	1.00	0.63	1.00	1.00	0.65	0.9
37.5		1.00	1.00	1.00	0.64	1.00	1.00	0.66	0.9
39		1.00	1.00	1.00	0.64	1.00	1.00	0.66	0.9
40.5		1.00	1.00	1.00	0.65	1.00	1.00	0.67	0.9
42		1.00	1.00	1.00	0.65	1.00	1.00	0.67	1.0
43.5		1.00	1.00	1.00	0.65	1.00	1.00	0.68	1.0
45		1.00	1.00	1.00	0.65	1.00	1.00	0.68	1.0
46.5		1.00	1.00	1.00	0.66	1.00	1.00	0.68	1.0
48		1.00	1.00	1.00	0.66	1.00	1.00	0.69	1.0
49.5			1.00	1.00		1.00	1.00	0.69	1.0
51			1.00	1.00		1.00	1.00	0.69	1.0
52.5			1.00	1.00		1.00	1.00	0.69	1.0
54			1.00	1.00		1.00	1.00	0.70	1.0
55.5			1.00	1.00		1.00	1.00	0.70	1.0
57			1.00	1.00		1.00	1.00	0.70	1.0
58.5			1.00	1.00		1.00	1.00	0.70	1.0
60			1.00	1.00		1.00	1.00	0.71	1.0
61.5			1.00	1.00		1.00	1.00	0.71	1.0
63			1.00	1.00		1.00	1.00	0.71	1.0
64.5				1.00			1.00		1.0
66				1.00			1.00		1.0
67.5				1.00			1.00		1.0
69				1.00			1.00		1.0
70.5				1.00			1.00		1.0
72				1.00			1.00		1.0
73.5				1.00			1.00		1.0
75				1.00			1.00		1.0
76.5				1.00			1.00		1.0
78				1.00			1.00		1.0
79.5				1.00			1.00		1.0
81				1.00			1.00		1.0

Table M16.2-8C Solid Sawn Timbers

Rating	1-HOUR				1.5-HOUR			2-HOUR	
Beam Width	5.5	7.5	9.5	11.5	7.5	9.5	11.5	9.5	11.5
Beam Depth	Design Load Ratio, R_s								
5.5	0.34	0.51	0.61	0.68	0.09	0.12	0.14		
7.5	0.51	0.77	0.92	1.00	0.32	0.45	0.54	0.15	0.20
9.5	0.61	0.92	1.00	1.00	0.45	0.64	0.76	0.32	0.43
11.5	0.68	1.00	1.00	1.00	0.54	0.76	0.91	0.43	0.58
13.5	0.72	1.00	1.00	1.00	0.60	0.85	1.00	0.51	0.68
15.5	0.76	1.00	1.00	1.00	0.64	0.91	1.00	0.56	0.76
17.5	0.78	1.00	1.00	1.00	0.68	0.96	1.00	0.61	0.82
19.5	0.80	1.00	1.00	1.00	0.70	1.00	1.00	0.64	0.87
21.5	0.82	1.00	1.00	1.00	0.73	1.00	1.00	0.67	0.91
23.5	0.83	1.00	1.00	1.00	0.75	1.00	1.00	0.70	0.94

Table M16.2-9 Design Load Ratios for Exposed Timber Decks

Double and Single Tongue & Groove Decking

(Structural Calculations at Standard Reference Conditions: $C_D = 1.0$, $C_M = 1.0$, $C_t = 1.0$, $C_i = 1.0$)

Rating	1-HOUR	1.5-HOUR	2-HOUR
Deck Thickness	Design Load Ratio, R_s		
2.5	0.22	-	-
3	0.46	0.08	-
3.5	0.67	0.23	0.03
4	0.86	0.40	0.12
4.5	1.00	0.56	0.25
5	1.00	0.71	0.38
5.5	1.00	0.85	0.51

Table M16.2-10 Design Load Ratios for Exposed Timber Decks

Butt-Joint Timber Decking

(Structural Calculations at Standard Reference Conditions: $C_D = 1.0$, $C_M = 1.0$, $C_t = 1.0$, $C_i = 1.0$)

Rating	1-HOUR				1.5-HOUR			2-HOUR	
Decking Width	5.5	7.5	9.5	11.5	7.5	9.5	11.5	9.5	11.5
Decking Depth	Design Load Ratio, R_s								
2.5	0.05	0.12	0.15	0.18	-	-	-	-	-
3	0.09	0.24	0.30	0.36	0.03	0.04	0.05	-	-
3.5	0.14	0.35	0.44	0.53	0.08	0.12	0.16	-	-
4	0.18	0.45	0.57	0.68	0.14	0.21	0.28	0.02	0.08
4.5	0.21	0.54	0.68	0.80	0.19	0.30	0.39	0.04	0.16
5	0.24	0.61	0.77	0.92	0.24	0.38	0.50	0.06	0.24
5.5	0.27	0.68	0.85	1.00	0.29	0.45	0.59	0.09	0.32

Example 16.2-1 Exposed Beam Example – Allowable Stress Design

Douglas-fir glulam beams span $L = 18'$ and are spaced at $s = 6'$. The design loads are $q_{live} = 100$ psf and $q_{dead} = 25$ psf. Timber decking nailed to the compression edge of the beams provides lateral bracing. Calculate the required section dimensions for a 1-hour fire resistance time.

For the structural design of the wood beam, calculate the maximum induced moment.

Calculate beam load:
$w_{total} = s (q_{dead} + q_{live}) = (6)(25 + 100) = 750$ plf

Calculate maximum induced moment:
$M_{max} = w_{total} L^2/8 = (750)(18)^2/8 = 30,375$ ft-lbs

Select a 6-3/4" x 13-1/2" 24F visually-graded Douglas-fir glulam beam with a tabulated bending stress, F_b, equal to 2,400 psi.

Calculate beam section modulus:
$S_s = bd^2/6 = (6.75)(13.5)^2/6 = 205.0$ in.^{3}

Calculate the adjusted allowable bending stress (assuming $C_D = 1.0$; $C_M = 1.0$; $C_t = 1.0$; $C_L = 1.0$; $C_V = 0.98$)
$F'_b = F_b (C_D)(C_M)(C_t)$(lesser of C_L or C_V)
$= 2,400 (1.0)(1.0)(1.0)(0.98) = 2,343$ psi   (*NDS* 5.3.1)

Calculate design resisting moment:
$M' = F'_b S_s = (2,343)(205.0)/12 = 40,032$ ft-lbs

Structural Check: M' ≥ M$_{max}$
40,032 ft-lbs ≥ 30,375 ft-lbs   OK

For the fire design of the wood beam, the loading i unchanged. Therefore, the maximum induced moment i unchanged. The fire resistance must be calculated. Calculate beam section modulus exposed on three sides
$S_f = (b – 2a)(d – a)^2/6 = (6.75 – 3.6)(13.5 – 1.8)^2/6 = 71.$ in.^{3}

Calculate the adjusted allowable bending stress (assuming C_D = N/A; C_M = N/A; C_t = N/A; C_L = 1.0; $C_V = 0.98$) $F'_b = F_b$ (lesser of C_L or C_V)
$= 2,400 (0.98) = 2,343$ psi   (*NDS* 5.3.1

Calculate strength resisting moment:
$M' = (2.85) F'_b S_f = (2.85)(2,343)(71.9)/12$
$= 40,010$ ft-lbs   (*NDS* 16.2.2

Fire Check: M' ≥ M$_{max}$
40,010 ft-lbs ≥ 30,375 ft-lbs   O

Design Aid
Calculate structural design load ratio:
$r_s = M_{max}/M' = 30,375/40,032 = 0.76$

Select the maximum design load ratio limit from Tabl M16.2-1B or calculate using the following equation:

$$R_s = \frac{2.85 S_f}{S_s C_D C_M C_t} = \frac{(2.85)(71.9)}{(205)(1.0)(1.0)(1.0)} = 1.00$$

Fire Check: R$_s$ ≥ r$_s$ 1.00 ≥ 0.76   O

Example 16.2-2   Exposed Column Example – Allowable Stress Design

A southern pine glulam column with an effective column length, $\ell_e = 168"$. The design loads are $P_{snow} = 16,000$ lbs and $P_{dead} = 6,000$ lbs. Calculate the required section dimensions for a 1-hour fire resistance time.

For the structural design of the wood column, calculate the maximum induced compression stress, f_c.

Calculate column load:
$P_{total} = P_{dead} + P_{snow} = 8,000 + 16,000 = 22,000$ lbs

Select a 8-1/2" x 9-5/8" Combination #48 southern pine glulam column with a tabulated compression parallel-to-grain stress, F_c, equal to 2,200 psi and a tabulated modulus of elasticity, E_{min}, equal to 880,000 psi.

Calculate column area:
$A_s = bd = (9.625)(8.5) = 81.81$ in.^{2}
$I_s = bd^3/12 = (9.625)(8.5)^3/12 = 492.6$ in.^{4}

Calculate the adjusted allowable compression stress (assuming $C_D = 1.15$; $C_M = 1.0$; $C_t = 1.0$):
$E_{min}' = E_{min} (C_M)(C_t) = 880,000 (1.0)(1.0)$
$= 880,000$ psi   (*NDS* 5.3.1

$F_{cE} = 0.822 E_{min}' / (\ell_e/d)^2$
$= 0.822 (880,000) / (168/8.5)^2 = 1,852$ psi (*NDS* 3.7.1.5

$F^*_c = F_c (C_D)(C_M)(C_t)$
$= 2,200 (1.15)(1.0)(1.0) = 2,530$ psi   (*NDS* 3.7.1.5

= 0.9 for structural glued laminated timbers (*NDS* 3.7.1.5)

$$C_P = \frac{1+F_{cE}/F_c^*}{2c} - \sqrt{\left[\frac{1+F_{cE}/F_c^*}{2c}\right]^2 - \frac{F_{cE}/F_c^*}{c}}$$

$$= \frac{1+0.7190}{2(0.9)} - \sqrt{\left(\frac{1+0.7190}{2(0.9)}\right)^2 - \frac{0.7190}{0.9}}$$

$$= 0.626 \qquad\qquad (NDS\ 3.7.1.5)$$

$F'_c = F_c^* (C_p) = 2,530 (0.626) = 1,583$ psi   (*NDS* 5.3.1)

Calculate the resisting column compression capacity:
$P' = F'_c A_s = (1,583)(81.81) = 129,469$ lbs

Structural Check: P' ≥ P_load
29,469 lbs ≥ 22,000 lbs OK

For the fire design of the wood column, the loading is unchanged. Therefore, the total load is unchanged. The fire resistance must be calculated.

Calculate column area, A, and moment of inertia, I, for column exposed on four sides:
$A_f = (b - 2a)(d - 2a) = (9.625 - 3.6)(8.5 - 3.6)$
$= 29.52$ in.^{2}
$I = (b - 2a)(d - 2a)^3/12 = (9.625 - 3.6)(8.5 - 3.6)^3/12$
$= 59.07$ in.^{4}
Calculate the adjusted allowable compression stress assuming C_D = N/A; C_M = N/A; C_t = N/A):

$F_{cE} = (2.03)\ 0.822\ E_{min}' / (\ell_e/d)^2$
$= (2.03)(0.822)(880,000) / (168/(8.5 - 3.6))^2$
$= 1,249$ psi ft-lbs   (*NDS* 16.2.2)

$F_c^* = (2.58)\ F_c = (2.58)(2,200)$
$= 5,676$ psi ft-lbs   (*NDS* 16.2.2)
$F_{cE}/F_c^* = 1,249/5,676 = 0.22$

$$C_P = \frac{1+0.22}{2(0.9)} - \sqrt{\left(\frac{1+0.22}{2(0.9)}\right)^2 - \frac{0.22}{0.9}} = 0.214$$

$F'_c = 5,676 (0.214) = 1,216$ psi

Calculate the resisting column compression capacity:
$P' = F'_c A_f = (1,216)(29.52) = 35,884$ lbs

Fire Check: P' ≥ P_load
35,884 lbs ≥ 22,000 lbs OK

Design Aid
Calculate structural design load ratio:
$r_s = M_{max}/M' = 22,000/128,043 = 0.17$

Select the maximum design load ratio (buckling) limit from Table M16.2-5A or calculate using the following equation:

$$R_s = \frac{2.03 I_f}{I_s C_M C_t} = \frac{(2.03)(59.07)}{(492.6)(1.0)(1.0)} = 0.24$$

Fire Check: R_s ≥ r_s   0.24 ≥ 0.17 OK

Example 16.2-3   Exposed Tension Member Example – Allowable Stress Design

Solid sawn Hem-Fir timbers used as heavy timber truss webs. The total design tension loads from a roof live and dead load are P_{total} = 3,500 lbs. Calculate the required section dimensions for a 1-hour fire resistance time.

For the structural design of the wood timber, calculate the maximum induced tension stress, f_t.

Calculate tension load: P_{total} = 3,500 lbs

Select a nominal 6x6 (5-1/2" x 5-1/2") Hem-Fir #2 Posts and Timbers grade with a tabulated tension stress, f_t, equal to 375 psi.

Calculate timber area: $A_s = bd = (5.5)(5.5) = 30.25$ in.^{2}

Calculate the adjusted allowable tension stress (assuming C_D = 1.25; C_M = 1.0; C_t = 1.0):

$F'_t = F_t (C_D)(C_M)(C_t)$
$= 375 (1.25)(1.0)(1.0) = 469$ psi   (*NDS* 4.3.1)

Calculate the resisting tension capacity:
$P' = F'_{c\ As} = (469)(30.25) = 13,038$ lbs

Structural Check: P' ≥ P_load
14,180 lbs ≥ 3,500 lbs OK

For the fire design of the timber tension member, the loading is unchanged. Therefore, the total load is unchanged. The fire resistance must be calculated.

Calculate tension member area, A, for member exposed on four sides:
$A_f = (b - 2a)(d - 2a) = (5.5 - 3.6)(5.5 - 3.6) = 3.61$ in.^{2}

Calculate the adjusted allowable tension stress (assuming C_D = N/A; C_M = N/A; C_t = N/A):

$F'_t = (2.85) F_t = (2.85)(375)$
$= 1,069$ psi ft-lbs   (*NDS* 16.2.2)

Calculate the resisting tension capacity:
$P' = F'_t A_f = (1,069)(3.61) = 3,858$ lbs

Fire Check: P' ≥ P$_{load}$ 3,858 lbs ≥ 3,500 lbs   OK

Design Aid
Calculate structural design load ratio:
$r_s = M_{max}/M' = 3,500/14,180 = 0.25$

Select the maximum design load ratio limit from Table M16.2-8C and divide by the load duration factor, C_D, o calculate using the following equation:

$$R_s = \frac{2.85 A_f}{A_s C_D C_M C_t} = \frac{(2.85)(3.61)}{(30.25)(1.25)(1.0)(1.0)} = 0.27$$

Fire Check: R$_s$ ≥ r$_s$ 0.27 ≥ 0.25   OK

Example 16.2-4   Exposed Deck Example – Allowable Stress Design

Hem-Fir tongue-and-groove timber decking spans L = 6'. A single layer of 3/4" sheathing is installed over the decking. The design loads are q_{live} = 40 psf and q_{dead} = 10 psf.

Calculate deck load:
$w_{total} = B(q_{dead} + q_{live}) = (5.5$ in./12 in./ft$)(50$ psf$) = 22.9$ plf

Calculate maximum induced moment:
$M_{max} = w_{total} L^2/8 = (22.9)(6)^2/8 = 103$ ft-lbs

Select nominal 3x6 (2-1/2" x 5-1/2") Hem-Fir Commercial decking with a tabulated bending stress, F_b, equal to 1,350 psi (already adjusted by C_r).

Calculate beam section modulus:
$S_s = bd^2/6 = (5.5)(2.5)^2/6 = 5.73$ in.^{3}

Calculate the adjusted allowable bending stress (assuming C_D = 1.0; C_M = 1.0; C_t = 1.0; C_F = 1.04):
$F'_b = F_b (C_D)(C_M)(C_t)(C_F)$
$= 1,350 (1.0)(1.0)(1.0)(1.04) = 1,404$ psi   (*NDS* 4.3.1)

Calculate resisting moment:
$M' = F'_b S_s = (1,404)(5.73)/12 = 670$ ft-lbs

Structural Check: M' ≥ M$_{max}$
670 ft-lbs ≥ 103 ft-lbs   OK

For the fire design of the timber deck, the loading i unchanged. Therefore, the maximum induced moment i unchanged. The fire resistance must be calculated.

Calculate beam section modulus exposed on one side:
$S_f = (b)(d - a)^2/6 = (5.5)(2.5 - 1.8)^2/6 = 0.45$ in.^{3}

Calculate the adjusted allowable bending stress (assuming C_D = N/A; C_M = N/A; C_t = N/A; C_F = 1.04):
$F'_b = F_b (C_F) = 1,350 (1.04) = 1,404$ psi

Calculate resisting moment:
$M' = (2.85) F_b S_f = (2.85)(1,404)(0.45)/12$
$= 150$ ft-lbs   (*NDS* 16.2.2

Fire Check: M' ≥ M$_{max}$ 150 ft-lbs ≥ 103 ft-lbs   OK

Design Aid
Calculate structural design load ratio:
$r_s = M_{max}/M' = 103/670 = 0.15$

Select the maximum design load ratio limit from Table M16.2-9 or calculate using the following equation:

$$R_s = \frac{2.85 S_f}{S_s C_D C_M C_t} = \frac{(2.85)(0.45)}{(5.73)(1.0)(1.0)(1.0)} = 0.22$$

Fire Check: R$_s$ ≥ r$_s$ 0.22 ≥ 0.15   OK

M16.3 Wood Connections

Where 1-hour fire endurance is required, connectors and fasteners must be protected from fire exposure by 1.5" of wood, fire-rated gypsum board, or any coating approved for a 1-hour rating. Typical details for commonly used fasteners and connectors in timber framing are shown in Figure M16.3-1 through Figure M16.3-6.

Figure M16.3-1 Beam to Column Connection – Connection Not Exposed to Fire

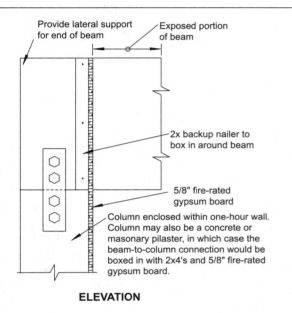

Provide lateral support for end of beam

Exposed portion of beam

2x backup nailer to box in around beam

5/8" fire-rated gypsum board

Column enclosed within one-hour wall. Column may also be a concrete or masonary pilaster, in which case the beam-to-column connection would be boxed in with 2x4's and 5/8" fire-rated gypsum board.

ELEVATION

Figure M16.3-2 Beam to Column Connection – Connection Exposed to Fire Where Appearance is a Factor

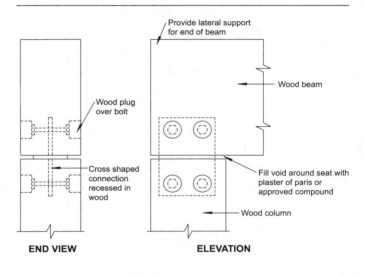

Provide lateral support for end of beam

Wood beam

Wood plug over bolt

Cross shaped connection recessed in wood

Fill void around seat with plaster of paris or approved compound

Wood column

END VIEW   ELEVATION

Figure M16.3-3 Ceiling Construction

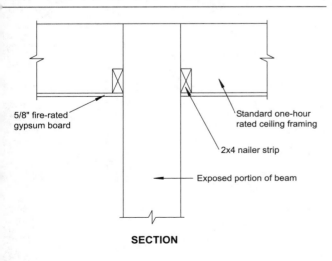

5/8" fire-rated gypsum board

Standard one-hour rated ceiling framing

2x4 nailer strip

Exposed portion of beam

SECTION

Figure M16.3-4 Beam to Column Connection – Connection Exposed to Fire Where Appearance is Not a Factor

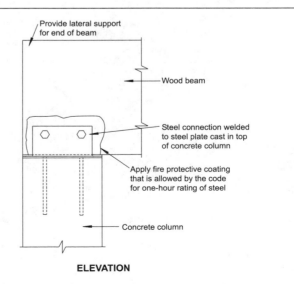

Provide lateral support for end of beam

Wood beam

Steel connection welded to steel plate cast in top of concrete column

Apply fire protective coating that is allowed by the code for one-hour rating of steel

Concrete column

ELEVATION

M16: FIRE DESIGN

16

Figure M16.3-5 Column Connections Covered

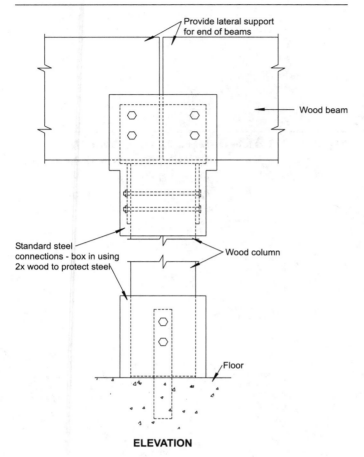

Provide lateral support for end of beams

Wood beam

Standard steel connections - box in using 2x wood to protect steel

Wood column

Floor

ELEVATION

Figure M16.3-6 Beam to Girder - Concealed Connection

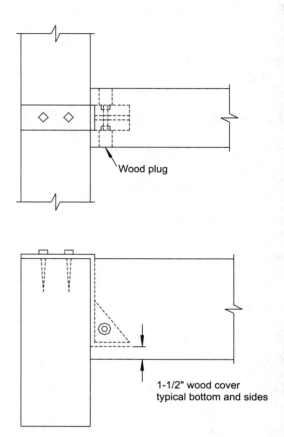

Wood plug

1-1/2" wood cover typical bottom and sides

ELEVATION